INTERNATIONAL AGRICULTURE

JAMES E. CHRISTIANSEN

Dr. Eddy Finley
Dr. Robert Price

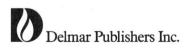
Delmar Publishers Inc.

NOTICE TO THE READER

Publisher does not warrant or guarantee any of the products described herein or perform any independent analysis in connection with any of the product information contained herein. Publisher does not assume, and expressly disclaims, any obligation to obtain and include information other than that provided to it by the manufacturer.

The reader is expressly warned to consider and adopt all safety precautions that might be indicated by the activities described herein and to avoid all potential hazards. By following the instructions contained herein, the reader willingly assumes all risks in connection with such instructions.

The publisher makes no representations or warranties of any kind, including but not limited to, the warranties of fitness for particular purpose or merchantability, nor are any such representations implied with respect to the material set forth herein, and the publisher takes no responsibility with respect to such material. The publisher shall not be liable for any special, consequential, or exemplary damages resulting, in whole or in part, from the readers' use of, or reliance upon, this material.

Cover Design: Brian Yacur

Cover Photo: Courtesy of NASA

DELMAR STAFF
New Product Acquisitions: Mark W. Huth
Senior Project Editor: Christopher Chien
Production Coordinator: Mary Ellen Black
Art Coordinator: Megan DeSantis
Senior Design Supervisor: Susan C. Mathews

For information, address Delmar Publishers Inc.
3 Columbia Circle, Box 15-015
Albany, New York 12203-5015

Copyright © 1994 by Delmar Publishers Inc.

All rights reserved. No part of this work covered by the copyright hereon may be reproduced or used in any form or by any means—graphic, electronic, or mechanical, including photocopying, recording, taping, or information storage and retrieval systems—without written permission of the publisher.

10 9 8 7 6 5 4 3 2 1 XXX 99 98 97 96 95 94

Printed in the United States of America
Published simultaneously in Canada
by Nelson Canada,
a division of The Thomson Corporation

Library of Congress Cataloging-in-Publication Data:

Finley, Eddy.
   International agriculture/by Eddy Finley and Robert Price.
     p. cm.
   Includes bibliographical references (p.   ) and index.
   ISBN 0-8273-5027-9
   1. Agriculture.   I. Price, Robert, 1910–   . II. Title.
S495.F56   1994
338.1—dc20   92-32760
                                 CIP

Table of Contents

SECTION I   INTERNATIONAL AGRICULTURE DESCRIBED   1

CHAPTER 1   Understanding the Importance of Agriculture to All Nations   3

Introduction *4*   Current Status of Agriculture in the U.S. *4*   Current Status of Global Agriculture *8*   International Agricultural Education and Extension Defined *10*   A Rationale for International Studies and Involvement *13*   Sharing the Global Harvest: A Kinship of Agriculturalists *14*   The Importance of Developmental Assistance *16*

CHAPTER 2   Population Growth and the Associated Food Deficit Problem   22

The Population/Consumption Explosion: A Challenge to Global Agriculture *23*   Overpopulation Defined *23*   Nature of Population Growth in an Agrarian Society *25*   Population Growth Rates and Concurrent Food Deficits *25*   Factors and Policies That May Result in Lowering of Birth Rates *26*   Population Figures Shown *27*   Migration of Farmers to the Metropolitan Areas *32*

CHAPTER 3   The Worldwide Problem of Providing for Adequate Human Nutrition   35

Concept of a Global Responsibility *36*   World Hunger Defined *37*   Malnutrition Defined *38*   The Hunger Cycle *41*   Broader Causes of Worldwide Hunger and Malnutrition *42*   Steps Nations Must Take to Overcome Malnutrition and Reduce Their Need for Imported Food *45*

CHAPTER 4   The Nature and Extent of Agricultural Production Worldwide   51

Introduction *52*   A Global Comparison of Selected Farm Types, Sizes, and Operations *52*   The Great Ranges in Productivity Among Nations *60*   The Development and Maintenance of Farming Systems of the World *62*   Nature and Extent of Farming Systems *65*   Extent and Distribution of Crops and Animals Providing the World's Food *66*   World Food Security and Overcoming Food Deficits *75*

CHAPTER 5   Environmental and Social Influences on Global Agricultural Production   80

Optimal Environmental Conditions *81*   The Effects of a Fragile Environment *81*   Increasing Severity of Environmental

Constraints *82*   The Task of Maintaining Optimal Environmental Conditions and of Lessening Constraints *89*   Sustainable Agricultural Production *90*   Social and Political Constraints on World Food Production *90*

CHAPTER 6   **Energy and Agricultural Production**   **94**

Classification and Uses of Energy Sources *95*   Energy Consumption by Food Systems in Developed Nations Compared to That in Developing Nations *97*   Sources of Power in Developing Nations *97*   Low Energy Efficiency and Corresponding Output in Developing Nations *98*   Energy-Intensive Compared to Labor-Intensive Economies *99*   Energy Use in United States Food Production *100*   Energy Reduction in Agriculture *100*   Alternative Energy in Agriculture *101*

CHAPTER 7   **Appropriate Agricultural Technologies**   **103**

Technology Defined *104*   Food Production and Technology *104*   Technology, Economic Growth, and Stability *107*   Technology and Organization for Rural Development *108*   Criteria for Evaluating Technologies Considered for Transfer *109*   Obtaining Practical Technical Expertise *111*

CHAPTER 8   **Dimensions and Implications of World Agricultural Trade**   **124**

Introduction *125*   Why Nations Trade with Each Other *125*   The Importance of Food Trade *126*   World Trade in Agricultural Products Compared to Production Locales *128*   Kinds and Types of Food Traded *129*   Actions and Policies used by Nations which Restrict Free Trade *135*   Agreements Between Nations which Determine World Trade *136*   National Benefits of World Agricultural Trade *138*   Aspects of World Trade That May Not Be Beneficial to Developing Nations *138*   Agricultural World Trade and the United States *140*   The Need for Agribusiness Involvement *143*   Staying Competitive in the Global Market *145*

CHAPTER 9   **Political Aspects of World Food Production**   **151**

The Effects of Politics on World Food Production *152*   Political and Politics Defined *152*   Policies Regarding Enforcement of Military and Police Powers *154*   Policies Regarding Food Acquisition and Distribution *156*   Policies Regarding Agricultural Production Incentives and Support *157*   Policies Regarding Use and Allocation of Resources *159*   Government Efforts toward Establishment and Maintenance of Support Systems *159*   U.S. Policies Regarding Assisting Developing Nations *160*

SECTION II — FOSTERING WORLDWIDE AGRICULTURAL DEVELOPMENT 163

CHAPTER 10 — Institutions and Agencies Providing Assistance and Training Programs 165

Concepts of Foreign Assistance Programs *166*   Types of Aid Given *166*   Classification of Agencies and Institutions Providing Aid *167*   Food and Agriculture Organization (FAO) of the United Nations *168*   United Nations Development Program (UNDP) *168*   Assistance by National Governments *172*   U.S. Assistance Programs *173*   Volunteer Agencies *177*   Other Organizations and Agencies *178*

CHAPTER 11 — Financing Programs for Agricultural Production and Development 182

Rural Finance—Requirements for an Effective System *183*   Farmers Require Financial Assistance *183*   International Programs of Financial Assistance *185*   Nature of U.S. Foreign Assistance *189*   International Agricultural Development Lending Institutions in the U.S. *190*

CHAPTER 12 — Human Resource Development—A Key to Development of Modern Agriculture 194

Human Resource Development *195*   The Change-Agent as Teacher *199*   The Adoption Process *200*   Maintenance of Incentives to Foster Production and Develop Enthusiasm *204*   Maintenance of a High Level of Constructive Participation *204*   The Outstanding Trainer/Change Agent *205*

CHAPTER 13 — Agricultural Education—Formal Instructional Services 209

The Nature of Society *210*   General Education Compared to Vocational Education *210*   Organization of Agricultural Education in Other Nations *212*   Primary, Elementary, and Secondary Schools of Developing Nations *214*   Agricultural Education in Paraguay: An Example of Success *216*   Relationship of Instruction at Primary and Secondary Levels to Agriculture in Developing Countries *217*   Institutions of Higher Education in Developing Countries *218*   Involvement of U.S. Universities *219*   Development Education for Citizens of Developed Countries *220*   Influences of Society on Education *221*

CHAPTER 14 — Agricultural Extension—Providing Instructional Services 225

The Importance of Agricultural Development and Training *226*   The Cooperative Agricultural Extension Service in the U.S. *228*   Organization of International Agricultural Extension Programs *229*   The Role of Extension Field Workers *231*   Women as Farmers and

Farm Laborers *231*   Rural Young People *238*   Elements Conducive to the Effectiveness of Agricultural Extension *238*   Legislative Provisions Needed *240*   Resources Needed *241*   Teaching Methods Adaptable for Education through Extension *241*   Selecting the Proper Teaching Methods *242*

CHAPTER 15 Agricultural Research—A Major Force in Development 247

Introduction *248*   Agricultural Experimentation in the U.S. *248*   U.S. Research System—Mission Oriented *249*   Agricultural Research in Other Nations *250*   Consultative Group on International Agricultural Research (CGIAR) *250*   The Nature and Scope of Agricultural Experimentation in Developing Countries *254*

CHAPTER 16 Design and Implementation of Agricultural and Rural Development 257

Introduction *258*   Agricultural and Rural Development Defined *258*   Program Planning for Agricultural Development *259*   Local Cooperatives as a Function of Agricultural and Community Development *262*

CHAPTER 17 Transforming Knowledge, Skills, and Commitment—Summing Up and Moving Forward 265

Introduction *266*   The Persistence of Nutritional Problems *266*   Positions Taken by Policy and Action Advocates of Agricultural Production *267*   Causative Factors Associated with Inadequate Agricultural Production *268*

APPENDICES 276

Appendix A	Gaining International Experience *276*
Appendix B	How to be an Effective Consultant on International Development Projects *296*
Appendix C	Agricultural Education: Definitions and Implications for International Development *301*
Appendix D	International Extension Programs for U.S. Citizens *307*
Appendix E	Role of the Land Grant University: An International Perspective *312*
Appendix F	Why Russia Can't Feed Itself *318*
Appendix G	Basic Principles for College and University Involvement in International Activities *322*

REFERENCES 326

INDEX 337

Preface

The United States is a member of an increasingly interdependent world community. Today, nations are linked by sophisticated communications, international financial and commodity markets, efficient and low-cost international transport systems, and both traditional and more innovative agricultural production systems. In many nations, such changes encourage development of highly skilled scientific institutions. Consequently, international concerns are everyone's concern—from the heartland of America, to big coastal cities such as Los Angeles and New York. Many employment opportunities will have an international dimension requiring one to three years in a foreign branch or subsidiary, or at least occasional overseas travel. An increasing proportion of U.S. students will either work abroad, work for a company that has vital interests abroad, or work for a company that experiences significant competition from abroad. Most importantly, all students will become citizens in a society which must find its way in an increasingly competitive and complex international environment.

Students must also understand the basic importance of agriculture to overall economic development, both in the U.S. and in other nations. Too often in the past, agriculture has been viewed as a tradition-bound sector with a sole mission of providing food and fiber. It is long overdue that agriculture be seen in a much broader context, and as the principal source from which overall development can rapidly and successfully emanate.

Agriculture in the U.S. cannot be readily understood unless there is a conscious effort to understand how it relates to those who are directly or indirectly associated with agriculture throughout the world. A study seeking to assist students to become knowledgeable of the relative significance of agriculture should have agricultural production worldwide as a central point of reference. However, most basic are the major aspects of processing, marketing, and distribution. It seems logical to include as a part of the study of international agriculture the particular influences that affect each of the aforementioned. For example, some of the influences include world population and the demand for food and fiber, energy and environment, appropriate technologies, social/cultural/political influences, world trade, financing of agricultural institutions, agencies providing supportive services, technology-transfer information, and literacy and resources for learning (as well as basic survival needs).

Because internationalization of the agricultural curriculum represents a most recent initiative for secondary and post-secondary curricular modification, this text was developed with the intent of infusing essential concepts and ideas into the subject matter.

There is strong evidence that improved education results in the development and well-being of a modern agriculture, both in the U.S. and abroad. Better educated and skilled people readily adopt new technology. People make the land productive and the industry successful. Investments in both basic and technical skills will greatly improve the potential for increasing agricultural output and all goods and services in a nation's economy.

"A shrinking world and increasing global interdependence are bringing new challenges and opportunities to everyone involved in agriculture. Organizations involved with international agricultural development efforts are placing increased emphasis on technical assistance and training related to improved and appropriate methods for the transfer of technology. A marked shift has also occurred in recent years, with the result that training programs in many countries are becoming more people oriented. Consequently, teachers of agriculture, extensionists, and related human development specialists are now sought for overseas development assignments on a scale commensurate with that experienced by agricultural scientists and technicians in recent years.

During the past two decades, agricultural institutions in the U.S. have been increasingly called upon to expand their participation in international agriculture activities, especially in the developing countries."* Therefore, portions of this text address some of the development activities in which various nations require assistance.

Ultimately, this text should serve to provide essential fundamental information as it pertains to agriculture internationally. Utilizing this text to enhance a student's awareness of the importance of international agriculture in general should bring about a greater appreciation of the role that agriculture plays in each nation's overall economy. It is hoped that upon completion of the text, the student will then be better prepared to begin studies within a particular international agricultural specialization, i.e., international agribusiness, agricultural finance, agricultural education, and many others.

Remaining abreast of world events is a difficult task because the one constant we have is change. For example, in March of 1990, Namibia won independence from South Africa. In May of that year, North Yemen and South Yemen united as one country. In July, Benin changed its flag. Currently, Croatia and Slovenia have seceded from Yugoslavia.

It should be further recognized that the Soviet Union accelerated movement toward "glasnost" (openness), thus, eleven of twelve former Soviet Republics proclaimed the birth of a democratic "Commonwealth of Independent States" and an end to the Soviet

---

*Source: Thummel, William L., Donald E. McCreight and Richard F. Welton. "The Role of Teacher Education in International Agriculture." In Arthur L. Berkey (Ed.), *Teacher Education in Agriculture,* 2nd Edition (p. 263), The Interstate Publishers and Printers, Danville, Illinois. Reprinted by permission of the American Association of Teacher Educators in Agriculture (1992).

Union. The approximately 284 million people in the eleven states are no longer Soviet citizens. They are citizens of their own independent states. Romania, Bulgaria, Czechoslovakia, Hungary, and Poland (collectively known as Eastern Europe) have assumed individual identities and are in the process of demanding democratic reform. Examples abound relative to the day-to-day changes in world events. It is essential that all students attempt to remain abreast of these events and the extent to which they are impacted by these events. Hopefully, this text reflects the current status of these evolving changes. It can be assured that "agriculture" will be a topic of study and discussion in every country and region of the world.

About the Authors

Eddy Finley spent his boyhood and adolescent years on a farm and ranch near Alanreed, in Donley County, Texas. He received his B.S. degree, with a major in Agricultural Education, from Texas Tech University in 1971, and his M.S. degree in Agricultural Education in 1976. In 1981, the Doctor of Education degree with a major in Agricultural Education was conferred by the Oklahoma State University.

While serving as agricultural instructor at Vernon Regional Junior College (Vernon, Texas), he was invited to join the faculty of Oklahoma State University in 1982 as an Assistant Professor and Coordinator of the Entry-Year Teacher Assistance Program. He currently serves in the capacity of Professor.

Dr. Finley was recognized by Alpha Zeta as the "Outstanding Professor" in the College of Agriculture in 1986 and in 1991 received the prestigious "Distinguished Teaching Award" awarded by Gamma Sigma Delta (the Honor Society of Agriculture). Also in 1991, Finley was recognized by OSU and the National Association of Colleges and Teachers of Agriculture for "Meritorious Teaching" in college. He is a recipient of the Honorary American Farmer Degree and has received the Distinguished Service Award from the National Vocational Agriculture Teachers Association. He has been named to *Who's Who in the South and Southwest, Who's Who in American Education,* and the *International Biographical Dictionary* (Cambridge, England). He is currently a member of the Board of Directors of the American Association of Agricultural Education (Southern Region), and has served as Theme Editor of The Agricultural Education Magazine. Finley is a member of Phi Kappa Phi Omicron Delta Kappa, Alpha Tau Alpha, Phi Delta Kappa, Gamma Sigma Delta, and Iota Lambda Sigma, as well as many other professional organizations.

He has sustained a strong interest in international agriculture, having served as academic advisor and instructor to undergraduate and graduate students from 26 other countries. Dr. Finley serves as a member of OSU's College of Agriculture International Programs Advisory Council. He also lived and worked in the Middle East for almost two years and has traveled extensively throughout the world.

Robert R. Price spent his boyhood and adolescent years on a farm near Oakwood, in Dewey County, Oklahoma. He received his B.S. degree, with a major in Horticulture, from Oklahoma State University in 1934, and his M.S. degree in Agricultural Education in 1946. In 1956, the Doctor of Education degree with a major in Agricultural Education was conferred by the Pennsylvania State University.

In 1948 he was invited to join the faculty of Oklahoma State University as an Assistant Professor and Itinerant Teacher Trainer in Agricultural Education. During twenty-seven subsequent years of tenure at the university, he held rank as Associate Professor, Professor, and Department Head, and beginning with retirement in 1975, Professor and Department Head Emeritus. In the years since, he has continued to be involved in student advisement, teaching, and writing.

Long a supporter and advocate of adult education and of community education, Dr. Price received the first Honorary Degree granted by the Young Farmers Association of Oklahoma. He also received recognition at state and national levels, being a recipient of the Honorary American Farmer Degree and the Distinguished Service Award from the Future Farmers of America Organization. Other recognitions came through activities in Oklahoma Adult and Continuing Education Association, Oklahoma Community Education Association, American Vocational Association, Gamma Sigma Delta, Phi Sigma, Phi Delta Kappa, and Phi Kappa Phi. In 1986, Dr. Price was honored by the institution he has served so many years in by induction into the Oklahoma State University Alumni Hall of Fame. Three years later, a similar honor was conferred by the Oklahoma Vocational-Technical Foundation with designation as one of the Cornerstones of Vocational-Technical Education in Oklahoma.

Dedicated to a strong conviction that vocational and technical education should be recognized as a most desirable component of international education, Dr. Price initiated a program of studies at OSU for agricultural education and extension workers in developing countries. He takes great pride in telling that during a period extending over thirty years, in which a total of 1120 international students completed study programs, he has served as major advisor for 214 students from 32 developing countries completing Master's and Doctoral degrees. Also, he has made brief consultative tours for the purpose of evaluating development programs to Saudi Arabia, Haiti, Bangladesh, and Papua, New Guinea.

Both authors are active in teaching international agriculture courses at OSU. Dr. Price, at the request of the OSU Independent and Correspondence Study Department, developed a university course on World Hunger and Agricultural Development. He authored, and later revised, a text-syllabus, "Education to Feed the World's Hungry," which is used in teaching the course. Dr. Finley teaches the course "International Programs in Agricultural Education and Extension" and has done so every semester since 1983. Since the second time the course was taught, it has filled to capacity with students every semester. It is a popular course among agriculture majors as well as nonagriculture majors.

The experiences of the authors, as well as internationally related materials reviewed covering a 30-year period, have culminated in the content presented within this text.

Acknowledgments

The authors wish to acknowledge the assistance given by many individuals during the preparation of this text:

Thomas C. Collins, Vice-President for Research and Dean of the Graduate College, Oklahoma State University, on behalf of the Patent Committee and the Board of Regents for Agricultural and Mechanical Colleges, Oklahoma State University, Stillwater, Oklahoma, for granting permission to use materials from Dr. Robert Price's study guide, "Foods Future: The Alleviation of World Hunger."

Karen Wagner, Director of Resource Development, Heifer Project International, Little Rock, Arkansas, for providing information and photographs as well as editing portions of this text.

James W. Cowan, Director, Division of Agriculture, National Association of State Universities and Land Grant Colleges (NASULGC), Washington, D.C., for granting permission to reprint portions of E. Boyd Wennergren and William Furlong's book, "Solving World Hunger: The U.S. Stake", and for granting permission to reprint "Basic Principles for College and University Involvement in Development Activities".

Ben Newcomer, Program Director, HEART Program, Warner Southern College, Lake Wales, Florida, for providing information and also for editing portions of this text.

William L. Miller, Executive Director, Midamerica International Agriculture Consortium (MIAC), Lincoln, Nebraska, for providing information requested, and for granting permission to reprint "How to be an Effective Consultant on International Development Projects".

Cathy Ringewald Cirina, Permission Supervisor, University of California Press, Berkeley, California, for granting permission to use material published in Walter Ebeling's book, "Fruited Plain: The Story of American Agriculture".

Dr. Theodore Hutchcroft, Program Officer/Communication, Winrock International, Morrilton, Arkansas, for providing permission to reprint materials published by Winrock International and for assisting with obtaining photographs.

Pedro Soto, Audio-visual Technician, and Donald Pohl, Director of Communication, CARE, New York, New York, for granting permission to reprint "Sharing the Global Harvest—A Kinship of Farmers" and for furnishing photographs.

J. Russell Bonel, Director, Division of Information, United Nations Development Program (UNDP), New York, New York, for providing information and for furnishing photographs.

Susan Fertig-Dyks, Director, Office of Publishing and Visual Communication, U.S. Department of Agriculture (USDA), Office of Public Affairs, Washington, D.C., for granting permission to reprint selected materials from various editions of the Yearbook of Agriculture published by the USDA, and for providing a list of available photographs.

Arlene W. Sullivan, Assistant to the Director, The Johns Hopkins University Press, Baltimore, Maryland, for granting permission to utilize selected materials from Sterling Wortman and Ralph Cumming's book, "To Feed This World".

Edgar Persons, Professor and Head, Department of Agricultural Education, University of Minnesota, St. Paul, Minnesota, for granting permission to reprint selected passages from "Looking at the World with a New Set of Glasses".

Marty Gibson, Mid-America Region Coordinator, Communications for Agriculture Exchange Program, Ames, Iowa, for providing requested information and for granting permission to reprint selected materials.

O. Donald Meaders, Professor Emeritus, Department of Agricultural and Extension Education, Michigan State University, East Lansing, Michigan, for granting permission to reprint materials from "Agricultural Education in Africa: A Systems View".

Carole Zimmerman, Director of Communications, Bread for the World Institute on Hunger & Development, Washington, D.C., for granting permission to utilize materials published by Bread for the World Institute on Hunger & Development.

Jane Sevier Johnson, Senior Writer-Editor, Publications Division, U.S. Agency for International Development (USAID), Washington, D.C., for granting permission to reprint materials from "USAID Highlights", "Frontlines", and "Agenda".

Clyde McNair, Photographer, "Front Lines", U.S. Agency for International Development (USAID), Washington, D.C., for forwarding requested materials and selecting photographs to be included in this text.

David Satter, author of "Why Russia Can't Feed Itself", and Katherine Burns, Editorial Rights and Permissions Manager, Reader's Digest, Pleasantville, New York, for granting permission to reprint the aforementioned article.

Paul Plawin, Editor, Vocational Education Journal, Alexandria, Virginia, for granting permission to reprint materials from Winifred I. Warnat's article, "World Class Work Force".

Evelyn Riley, on behalf of Margaret M. Patch, Permissions Administrator, Addison-Wesley Publishing Company, for granting permission to reprint materials from Les Donaldson and Edward Scannell's book, "Human Resource Development: The New Trainers Guide".

Keith Richmond, Chief Editor, Food and Agriculture Organization (FAO) of the United Nations, Rome, Italy, for granting permission to utilize materials from "Agricultural Extension—A Reference Manual", 2nd Edition, and for providing photographs to be used in this text.

Suzanne C. Toton, Professor, Department of Religious Studies, Villanova University, Villanova, Pennsylvania, for granting permission to reprint selected portions of her book, "World Hunger: The Responsibility of Christian Education".

Doris Goodnough, Right & Permissions, Orbis Books, Maryknoll, New York.

Bob R. Stewart, President, American Association for Agricultural Education (on behalf of the AATEA and Dr. Arthur Berkey, Editor), c/o Agricultural Education Department, University of Missouri, Columbia, Missouri, for granting permission to reprint materials from the book, "Teacher Education in Agriculture".

Ronald L. McDaniel, Vice President-Editorial, Interstate Publishers, Inc., Danville, Illinois, for granting permission to reprint selected materials from Robert S. Spitzer's book, "No Need for Hunger".

Jack C. Everly, Editor, NACTA Journal, Urbana, Illinois, for granting permission to reprint D. Craig Anderson's article, "Agricultural Education: Definition and Implications for International Development", Richard Welton's article, "Agricultural Education in Paraguay", and James E. Diamond's article, "Philosophy of International Agricultural Development".

G. Edward Schuh, Dean and Professor, Hubert H. Humphrey Institute of Public Affairs, Minneapolis, Minnesota (Note: Dr. Schuh is former Director, Agriculture and Rural Development of The World Bank), for granting permission to reprint "International Extension Programs for U.S. Citizens" and "Role of the Land Grant University: An International Perspective".

Moya, Permissions Department, Friendship Press, New York, New York, for granting permission to use materials from Phillip Wogaman and Paul McCleary's book, "Quality of Life in a Global Society", and Larry Minear's book, "New Hope for the Hungry".

Toby Baker, Project Administrator, DeKalb Plant Genetics, DeKalb, Illinois, for obtaining photographs suitable for this text.

E. Boyd Wennergren, Professor, Agricultural Economics, Utah State University, Logan, Utah, for his personal interest in this text and for the assistance provided in obtaining permission to reprint materials from "The United States and World Agriculture Development", which he and William Furlong authored.

Anatasia Broderick, Manager Library Services and Permissions, for granting permission to utilize excerpts from Sue Rodwell Williams' book, "Nutrition and Diet Therapy", C. V. Mosby Company, St. Louis, Missouri.

B. Glass, Director of Marketing, for granting permission to utilize excerpts from John Maxwell Hamilton's book, "Main Street America and the Third World", Seven Locks Press, Cabin John, Maryland.

Betty J. Evans, Permissions, for granting permission to utilize excerpts from the National Geographic Society's book, "Our World–National Geographic Picture Atlas of Our World", Washington, D.C.

Martha L. Peacock, Managing Editor, for granting permission to utilize excerpts from Michael F. Lofchies book, "The Policy Factor: Agricultural Performance in Tanzania and Kenya", Lynne Rienner Publishers, Inc., Boulder, Colorado.

Norma J. Johnson, Rights and Permissions Coordinator, for granting permission to utilize excerpts from Allen D. Jedlicka's book, "Organization for Rural Development", Praeger Publishers, Greenwood Publishing Group, Westport, Connecticut.

Donnella H. Meadows, for granting permission to reprint her article titled, "If the World were a Village of 1,000 People", The Old Farmers Almanac, Yankee Publishing Company, Dublin, New Hampshire.

The authors are deeply indebted to the aforementioned for their interest, cooperation, and support regarding the development of this text.

Further appreciation is extended to the following institutions and agencies that provided photographs, charts, and graphs in order to supplement this text:

Winrock International, Morrilton, Arkansas.
United States Agency for International Development (USAID), Washington, D.C.
Heifer Project International, Little Rock, Arkansas.
CARE, New York, New York.
United States Department of Agriculture (USDA), Washington, D.C.
United Nations Development Program (UNDP), New York, New York.
Food and Agriculture Organization (FAO) of the United Nations, Rome, Italy.
DeKalb Plant Genetics, DeKalb, Illinois.
National Association of State Universities and Land Grant Colleges (NASULGC), Washington, D.C.
Consortium for International Cooperation in Higher Education, Washington, D.C.

The authors also wish to acknowledge the following individuals who reviewed the manuscript and provided additional insight, thus contributing greatly to the continuity and relevance of this text:

James E. Christiansen, Professor, Department of Agricultural Education, Texas A & M University, College Station, Texas.
O. Donald Meaders, Professor Emeritus, Department of Agricultural and Extension Education, Michigan State University, East Lansing, Michigan.
Kary Mathis, Professor and Chair, Department of Agricultural Economics, Texas Tech University, Lubbock, Texas.
Edgar Persons, Professor and Head, Department of Agricultural Education, University of Minnesota, St. Paul, Minnesota.
Theodore Hutchcroft, Program Officer/Communication, Winrock International, Morrilton, Arkansas.
Phillips Foster, Professor, Department of Agricultural and Resource Economics, University of Maryland, College Park, Maryland.

Further appreciation is expressed to:

Dr. William Weeks, Assistant Professor, Department of Agricultural Education, Oklahoma State University, for the computer-generated graphics he skillfully developed.

Dr. H. Robert Terry, Professor and Head, Department of Agricultural Education, Oklahoma State University, for his continued moral support and encouragement.

Mrs. Tresa Runyan, Secretary, Department of Agricultural Education, Oklahoma State University, for her computer expertise, assistance, and moral support.

Finally the authors most gratefully acknowledge their indebtedness to the many international students from some 49 different nations who shared so enthusiastically through class discussion, research, reports, and personal conferences. This text has been greatly enhanced through the contributions of more than 800 of those agricultural professionals.

The cooperation given by publishers and various periodicals and books named either in the credits or references is greatly appreciated.

SECTION 1
International Agriculture Described

CHAPTER 1
Understanding the Importance of Agriculture to All Nations

Source: United Nations Development Program; Photo by John Issac.

Terms to Know

technology transfer techniques
biological research
underdeveloped (developing) countries
emerging countries
agricultural and food policy
public policy research
marketing
Third World
First World
Second World
agricultural education
agricultural extension
agricultural development
economic development
developmental assistance
appropriate technology
Food for Peace Program

Objectives

After reading this chapter, you should be able to:

- discuss the important contributions of agriculture to a country's development.
- discuss the meaning of agricultural education and extension, internationally.
- explain why we are truly part of an international system and why the international system impacts each of us.
- compare the similarities and differences between a developed and a developing country.
- explain why agricultural development in other countries is important to our own country.

Introduction

Paramount to any discussion of the means by which people of the world may experience more meaningful and productive lives is a more knowledgeable understanding of the role played, and the contributions made, by the industry of agriculture. Such understanding is needed not only by those engaged in work directly related to agriculture, but to all people, of all nations.

Decisions made by every country regarding the allocation of resources to their agricultural industry need to be based upon accurate knowledge and understanding.

Agriculture is a business, not only for each individual production unit, but for each unit engaged in processing, distribution, and marketing as well. Obtaining knowledge and skills is essential globally in every phase of agriculture.

Although somewhat primitive agricultural operations still exist in many of the world's nations, sophisticated technical and research operations characterize many of the more developed nations. People living in these nations generally enjoy a much higher standard of living.

Because agriculture in the U.S. is representative of approximately 20 to 30 nations achieving and sustaining highly successful programs, let us briefly examine the current status of U.S. agriculture in order that comparisons may be made.

Current Status of Agriculture in the United States

The United States has abundant natural resources and excellent technology to help lead the world in agriculture. However, the richest resource of any nation's agricultural industry is its people. In the U.S., over 20 million men and women drive the productivity of the U.S. food and agriculture industry.

They strive to seek out the best crop varieties, select the healthiest livestock, and find the best packaging for cut flowers. They develop a system to grow food in space. They look for ways to cure or prevent diseases in plants, animals, and humans. They find exciting new ways to teach young people as well as adults.

The initiative, creativity, and hard work of these people promote U.S. efficiency and its ability to compete in the world marketplace. However, it is perhaps unavoidable that in today's complex society, many people in the U.S. still hold a simplistic view of agriculture—a barn and silo along the road, planting and harvesting a crop, or getting up early to milk the cows.

The American food and agriculture industry today is this and much more. It encompasses thousands of agribusiness firms that process, deliver, and sell food and other products to domestic and global customers. It involves institutions that provide credit, machinery, and information; scientists who contribute to greater bounty with less environmental impact; and people who teach and study agriculture.

American farmers produce a hundred or more different crops and husband dozens of breeds of livestock on land that ranges from lush pasture lands to deserts, plains to mountains, and vast rural areas to suburbia. U.S. farm families spring from differing ethnic backgrounds and work successfully in every kind of farming endeavor.

The U.S. field of agriculture abounds with excellent local, regional, national, and international opportunities: exciting careers to help sustain international competitiveness; enhanced marketing opportunities in a global trading environment; fostering efforts to ensure maintenance of an abundant and healthy food supply; searching for additional industrial uses for farm products; and producing food and fiber in an environmentally sensitive way while still providing a fair return for farmers and others.

Figure 1–1 depicts the food and fiber system, as well as the industries essential to agricultural production. Input industries include manufacturing, mining, trade, finance, insurance, and government. Output industries include processing and marketing, wholesale and retail trade, and indirect agribusiness. Agricultural production is comprised of farming, agricultural services, forestry, and fisheries. Also emphasized is the interrelationship between developed and developing nations.

It is critical to understand the diversity of domestic and global agriculture and how it influences our daily lives—whether or not we work in agriculture.

U.S. agricultural programs provide essential technical and managerial skills that people need in order to succeed. Agricultural education, extension, and research programs provide the up-to-date technology and information farmers need to increase production and farm incomes. Agricultural youth programs develop the leadership and organizational skills required to promote agriculture and improve standards of living.

Many other nations worldwide have followed a similar pattern by investing in agricultural programs and incorporating **technology transfer techniques** much like those of the United States. These human resource investments are now paying important dividends in the form of increased agricultural productivity. The process of developing, transferring, and using agricultural technology requires knowledgeable people at all levels. A nation must provide appropriate agricultural programs if it is to be successful in developing its agricultural potential in rural areas and in meeting its national food requirements.

Agricultural processes that transform resources into products for consumers are constrained by human and natural resources: economics, germ-plasm for crop varieties and animal breeds, and diseases and insects. The history of mankind is largely a struggle to reduce these constraints. Our natural resources are finite, while mankind's capacity to want more foods, goods, and services is infinite. Science offers alternatives to make limited resources go farther to produce food. Some of these means, such as technology and production practices, are revolutionary; others are modest contributions. Whether modest or revolutionary, worldwide improvements will not be forthcoming without foresight and preparation.

Agriculture provides products for basic human needs of food, fiber, and shelter. It also provides inputs used by virtually all sectors producing goods and services in our economy. Because resources cannot supply all wants for agriculture and other products, the U.S. relies on the marketplace to determine:

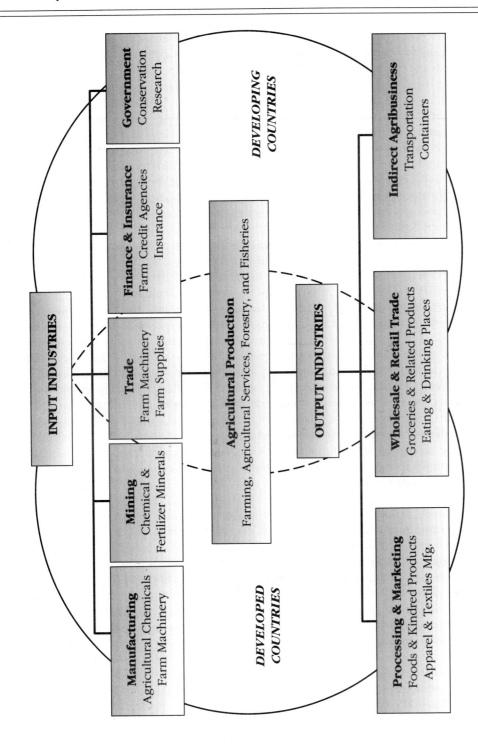

Figure 1–1 • The Food and Fiber System. *Source: United States Department of Agriculture.*

- the level and mix of resources and products.
- the methods of producing and processing.
- allocation of products over geographic space and time.

Generally speaking, export markets are a source of economic instability. Marketing decisions of farmers and ranchers have become more important and complex due to factors such as increased price volatility. Marketing tools such as market information, market projections, and commodity futures are increasingly being used to reduce marketing uncertainty both in the U.S. and abroad.

U.S. farmers have utilized technology to increase food and fiber output per acre, per person hour of labor, and per animal. Though energy is a critical component of this technology, less than 3.0 percent of it consumed in the U.S. is used for production of raw agricultural products. Timeliness is essential in many farm operations that require energy, and the flow of energy supplies when they are needed must be assured.

Much of the agricultural progress in the U.S. can be traced to basic research. Fundamental **biological research** is concerned with the structure and function of biological systems such as how cells differentiate into tissues and organs, how organisms sense their environment, and how organisms transmit genetic information. It has been taken for granted that fundamental biological research will continue to supply the knowledge base for applying practical solutions to emerging problems; however, our basic stock of knowledge is being depleted. Fundamental biological research will need to be stepped up to ensure that applied research can draw from a basic pool of knowledge to turn out productive and profitable practices for farms and ranches here in the U.S. and abroad.

Historically, real prices of agricultural products have increased less than prices of nonfarm goods and services. Between 1950 and 1970, the Consumer Price Index increased at 2.5 percent per year while farm prices increased by 0.3 percent per year. During this period and indeed for most of this century, the supply of agricultural products grew faster than demand. Because the U.S. had reached a standard of living such that a typical family spent additional money on cars and TV sets rather than on food as real income increased, domestic demand growth was determined more by growth in population than in real income.

Projecting the volatile export component of worldwide agricultural demand is a challenge. Demand for our agricultural exports is determined by growth in worldwide population, income, agricultural output, and value of the dollar, as well as by political considerations. (Weather is transitory and its effect on exports averages out in the long run.) The major growth in U.S. export demand will come from **underdeveloped** or **developing countries,** and their ability to import will be tied to the general economic conditions that prevail over the next two decades. Additional growth in U.S. export demand should also come from **emerging countries** in Eastern Europe. These emerging countries, for example, include Bulgaria, Czechoslovakia, Hungary, Poland, Romania, and Yugoslavia.

Projecting the future rate of increase in aggregate agricultural supply worldwide is nearly as uncertain as projecting export growth; however, it is essential that such projections be made in order that there is an accounting of food supply needs as well as anticipated food supply reserves.

Affordable computers in the 1980s have changed U.S. farm management procedures to an extent comparable to that experienced in the transition from animal power to machines in the past 50 years. In the past, many of the existing techniques of analysis and planning have been limited to firms large enough to afford expensive computers; however, there is an ever increasing number of U.S. farmers who are taking advantage of personal computers and software to assist them in their management endeavors. Just as importantly, each nation worldwide has access to sophisticated computers to assist them in analyzing and planning their national and international agricultural programs.

In the U.S., public policy directly affects not only the economic well-being of the nation's farmers, but the economic well-being of every country as well. Decisions made by government often impact farm income as much as do management and marketing decisions made on the farm.

Agricultural and food policy encompasses a broad range of policies including land-use and conservation policies, marketing policy for both input and product markets, foreign trade policy, nutrition and food distribution policies, and commodity price and income support policies. Because of the many decisions that have to be made regarding the aforementioned policies, **public policy research** is an important method of determining recommendations and solutions. The purpose of public policy research (and educational programs) is to provide information on options and their impacts so that legislators, farmers, and others can make decisions.

The U.S. has the capabilities to produce an abundance of food and fiber. Just as importantly, the U.S. has strong marketing capabilities (thanks to energetic, innovative and enlightened, responsible organizations and governmental policies). Processing, storing, transporting, and exchanging of agricultural commodities are all parts of a broader category called **marketing.** Marketing includes all of the physical and coordinative activities linking producers of agricultural commodities with consumers of finished products. Pricing, exchange, and coordinative activities are functions performed by markets within a marketing system. Markets allocate available supplies among consumers, reward producers of highly valued products, and motivate producers to allocate resources to their most efficient use. For these reasons, marketing is sometimes referred to as the important "other half" of agriculture. An efficient agricultural marketing system is essential to the economic viability of the production half of agriculture.

Current Status of Global Agriculture

Today, less than a dozen nations produce ample food for their own people, approximately 50 nations are essentially food self-sufficient, and the rest (or nearly 120 nations) are dependent on outside food sources. Of these latter 120 nations, most are considered to be economically underdeveloped. Figure 1–2 outlines and distinguishes between the developed and developing regions of the world.

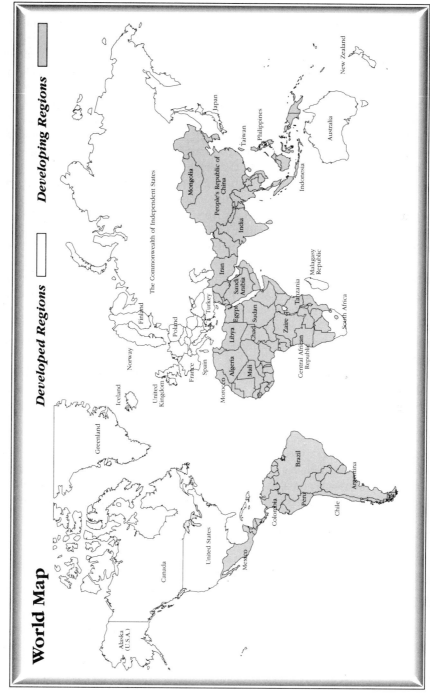

Figure 1–2 • World map. Source: *National Association of State Universities and Land Grant Colleges.*

So, what is a developing or **Third World** country? According to Hamilton (1986), "developing (or underdeveloped or less-developed) countries organized themselves into the "Group of 77," the Third World economic caucus that really includes more than 100 countries. It is India with 750 million people and Fiji with 700,000. It is Communist China and conservative Singapore. It is Burkina Faso and Bangladesh, with annual per capita incomes of less than $200, and it is newly industrialized countries like Brazil and Seychelles, with per capita incomes of $2,000. It is oil rich Saudi Arabia, Venezuela, and Nigeria, and oil poor Tanzania and South Korea. It is also, for example, Turkey, Chile, and Costa Rica, Sudan, and Senegal, Thailand, and Taiwan, and Yugoslavia. Attempting to define developing country is too complex for neat definitions; however, it is too big to be ignored."

Experts attempt to bring precision to the definition of developing country. Others simply talk in terms of North and South, suggesting inaccurately that developing countries lie strictly in tropical regions, while industrial nations lie in the temperate northern climates. (Furthermore, the term **First World** is today used to describe industrialized nations. Eastern bloc countries in Europe are sometimes placed in the **Second World**.) Nearly 90 percent of the people in these developing countries live in rural areas. For many, maintaining even a subsistence-level lifestyle is a daily concern. Many international and national organizations have attempted to help these rural families by increasing their agricultural output with the anticipation that increased agricultural productivity would enhance the nation's overall economy, causing them to become potential customers for U.S. products; see Figure 1–3. Because of the large number of these nations it is imperative that we study their impact on global agriculture supply and demand as well.

One way in which to bolster agricultural production is by developing educational (formal and nonformal) and research systems geared toward people within the agricultural sector; see Figure 1–4. Developed nations have worked hard to develop adequate agricultural programs and many developing nations are working hard to develop them also. However, in some cases, the importance of agriculture is still poorly understood and is frequently given low priority. At the local level for example, to the 60 to 80 percent of Africans living in small villages, development means such things as children in school, nearby outdoor water taps, a small mill to help grind maize, and the maintenance of good soil fertility. In other cases, the importance of human resource development is recognized, but nations lack agricultural educators, extension personnel, and researchers who can plan and implement these programs at all levels.

International Agricultural Education and Extension Defined

It is important to address the role of agricultural education and extension, because if it is to be of relevance to the problems that our society and others face: it has to address our society in the dimensions in which it actually exists. We now live in a truly interdependent economy, one in which developments in other countries are as important as

Figure 1–3 • In Nepal, a CARE agricultural extension worker, left, teaches a villager improved agricultural methods. Nepal is one of more than 30 developing nations where CARE operates programs to help impoverished people use their land and resources wisely so the environment—and their own livelihood—can be preserved and enhanced. *Source: CARE; photo by Jon Burbank.*

developments in our own economy (Schuh, 1985). To make informed and sensible decisions in today's world, we have to understand the world as it is, not as it used to be. The technological advances in communications, transportation, and information have changed our world forever.

According to Anderson (1984), **agricultural education,** in the context of international agriculture, embraces a wide range of meanings, for it is any organized activity that has as its purpose instruction in agriculture. For example, in other countries agricultural studies in the disciplines of Animal Science, Agronomy, and Agricultural Economics are referred to as studies in agricultural education. Also, with exception to the United States, agricultural education in other countries generally is accepted as being synonymous with cooperative extension. Meaders, et al. (1988) asserted that agricultural education is characterized by having an interface with both education and agriculture. Whether dealing with policy at the national level or dealing with students in a particular classroom, the implications are for interface with both education and agriculture. The nature of the goals, at whatever level, should involve both the educational and agricultural systems. The nature of the goals may be both for societal needs and individual needs.

Figure 1–4 ● A teacher at the San Narciso REAP school in Belize teaches agricultural science through student participation. REAP—Relevant Education for Agriculture and Production—is a project designed by CARE and the Ministry of Education to teach agricultural production to rural primary and secondary students. The program, which began in 1977, has since been expanded to include urban students. Instructors adapt agricultural activities to an urban setting and help teach students employment training, manufacturing, construction, marketing, and distribution. *Source: CARE; photo by Silvano Guerrero.*

For purposes of this text, international agricultural education and extension is any organized activity, both formal and informal, which has as its purpose instruction in and about agriculture interfaced with educational programs, business, and industry. Agricultural education is truly integrated.

Swanson (1984) specifically defined **agricultural extension** in AGRICULTURAL EXTENSION, A REFERENCE MANUAL, published by the Food and Agriculture Organization of the United Nations, as being an on-going process of getting useful information to people (the communication dimension) and assisting those people to acquire the necessary knowledge, skills, and attitudes to effectively utilize this information or technology (the educational dimension). Generally, the goal of the extension process is to enable people to use these skills, knowledge, and information to improve the quality of life.

Agriculture is complex and easily misunderstood by many people. It is a way of life for some; it is a business, profession, or science, and just another sector of the economy to others. (To many consumers, it may simply be the source of most of their food supplies—resulting in a short trip to the supermarket.) Because agriculture is essential to the lives of all human beings, it is imperative that food supplies be increased to meet the needs of an expanding population. Because productive and profitable farming can contribute to higher standards of living sought so widely, the nature of agriculture and the requirements for its improvements need to be widely understood.

A Rationale for International Studies and Involvement

As stated by Schuh (1987), large continental countries like the United States tend to be insular and inward-looking. They tend to be self-contained, and to have little dependence on the rest of the world. Because they believe they don't need the rest of the world, they tend to ignore it. Schuh (1987) further suggested that most of the U.S. students will either work abroad, work for a company that has vital interests abroad, or work for a company that experiences significant competition from abroad. Alternatively, they will work in the public sector, large parts of which also are engaged with the global economy in one form or another. Most importantly, all of our students will be citizens in a society which must find its way in an increasingly competitive and complex international economy. They will need to elect representatives who can make correct decisions in this kind of world—decisions that will affect their well-being far into the future. Perhaps the most important point to be made here is that we are now part of an international system, and thinking about ourselves as a self-standing, independent economy is no longer relevant.

Obviously, none of us can understand everything. But we can acquire knowledge about those parts of the world that are in our vital interests. If you are a soybean producer, for example, you need to understand how the international system works, as well as which are the major consuming and producing countries. It is sometimes argued that a developing country that expands its own agricultural productivity will be able to reduce its imports from the United States and perhaps even compete with U.S. farmers for markets in other countries. U.S. farmers may see this as a threat to their traditional markets and perhaps to future export expansion. On its face, that argument appears to be logical, and in fact, there are some specific cases where agricultural improvement in a given country may have caused the U.S. to lose market share with a specific customer for a specific product temporarily.

In most of the cases in which **agricultural development** has helped another country increase its income, it has also caused that market to expand as a customer for U.S. farm products. The evidence is overwhelming that countries in that situation will most often experience a rapid growth in demand. As they succeed in expanding food production, the expectations of their people are enlarged, and a substantial share of the new income generated will be spent for additional imports of food.

Many economists, farm and commodity organization leaders, and policy makers profess and firmly believe American farmers have little to fear and much to gain from agricultural improvement in either developing or emerging countries. Such development in rural sectors is essential if those countries are to expand incomes, consumption, and imports of agricultural products, including food. This fact alone offers U.S. agriculture its greatest opportunity to resume growth in the export market in the years to come.

What most Americans need to realize is that it took up until 1950 to populate the earth with 2.5 billion people. Then by mid-1987, we had added another 2.5 billion people—in less than 40 years. Presently, there are about 5.2 billion people, and even though birth rates have slowed significantly, demographics dictate that it will still be some time before population growth stabilizes (Yeutter, 1989).

Currently, almost a billion people are added to the earth's population every decade. That is equivalent to one more China or India, or four more Americas to feed every 10 years.

By the year 2025, it is projected that there will be about 8.4 billion people on earth, competing for the same products of the earth's natural resource base. Only with the rational application of knowledge and science, combined with economic and environmental decisions, can we enhance productivity to the level that will soon be needed.

To add even more emphasis to this situation, 95 percent of those additional 3.2 billion people will be living in the less developed nations where major problems such as lack of food, severe poverty, illiteracy, and over-population exist.

Sharing the Global Harvest: A Kinship of Agriculturalists*

We believe we can resolve the paradox of the food-poverty-population-illiteracy problem in the midst of planning. The solutions must come from agriculturalists willing to devote time and energy to this important international concern. It is a challenge not to be overlooked. We need to better understand the role agriculture plays in this process at home and in the developing countries.

It is evident that in the last 100 years, U.S. agriculture has grown from subsistence to abundance. We must share the lessons we have learned with agriculturalists around the world. We must be able to function within this complex world and to be able to better understand and interact with our world neighbors in both developed and developing countries.

Historical evidence points to the beginnings of plant cultivation and animal husbandry about 10,000 years ago. Through the centuries, the simple techniques of crop cultivation and animal husbandry remained unchanged. Almost everybody worked at food production.

---

*Portions of this have been adapted from CARE, New York, NY.

Not until the last 100 years did the results of scientific inquiry begin to transform agriculture. It should be recognized that agriculture in America is the backbone of this nation. A nation which grew from modest and difficult beginnings, America has grown to become among the leaders in world agriculture. To help understand the efforts to move beyond poverty and toward economic freedom in the more than 100 less developed nations known as the Third World, we should examine our own beginnings. Efforts alone could not assure meeting the goals of the American farmer. It could not guarantee sound market systems to accept farmer's products nor could it promise a decent return on their investment and toil.

America's early days were marked by the struggle to grow food. The battle with the elements was fought with few, if any, defenses. Our farmers faced tough marketing problems, lack of credit, and sometimes shrewd lenders and middlemen who bankrupted many and plunged others into debt. American farmers learned through trial and error and invention. They learned to utilize opportunities: the land's basic richness, the Homestead Act's incentives of acreage to those willing to farm it, the services of the Land Grant Colleges and Agricultural Education and Extension Offices, and the newly forming market systems.

American farmers endured many economic and political crises, including the Great Depression and Dust Bowl era of the 1930s, when suddenly poverty in America was common place.

Today, agriculturalists in Asia, Africa, and Latin America are trying to cope with many of the same problems in their own situations; but, are fighting poverty more prolonged and severe than anything Americans have experienced.

In this country, agriculturalists sought to solve some of their problems by joining hands and forming their own organizations. People fought for farm credit administration, for insurance, and for other measures in order to receive fair prices. They realized they had to create these opportunities and institutions. Even the strongest individual effort could not overcome financial uncertainty or influence farm policies. In addition to the Homestead Act, others such as the Morrill Act, Hatch Act, Smith-Lever Act, and Smith-Hughes Act, which established the Land Grant Colleges, Agricultural Experiment Stations, Cooperative Extension Service, and Vocational Agriculture programs, respectively, provided support to the development of agriculture in the U.S. Other organizations in the form of cooperatives, such as the Rural Electric Cooperatives, contributed greatly to the advancement of rural people.

Today, agriculturalists in other nations, especially the developing nations, are trying to rise above formidable constraints and make faster progress with what means they have. What they envision, as our agriculturists did, is greater influence over their own lives and a better chance to produce. Many agriculturalists worldwide are just now building and strengthening their position in order to expand the importance of agriculture in their own national economy.

Our experiences in America illustrates some of the key factors to be considered by these countries and their agriculturalists. Some of these experiences include the crucial role of incentives to encourage producers and the need for an efficient marketing system.

Agriculturalists worldwide are not only tackling the problems we faced as a young country, but are also confronting those we face today, including volatile inflation and interest rates, complex world market forces, low commodity prices, loss of topsoil, and reliance on an environment which turns hostile with drought, insects, fire, disease, or flood.

Many of our neighboring countries are young (having recently gained independence), mostly rural nations that are experiencing tremendous growing pains. While we may be concerned with the consequences of underdemand and oversupply, our world neighbors may just be trying to feed their families. While we keep an anxious eye on our foreign markets, they may be seeking credit for the first time, or applying a technology new to them to do better in the local market. They are trying to build themselves from the ground up as we did, but in a more hazardous local environment and in an economically complex world.

No one is suggesting that our history as a developing nation and the histories of today's developing countries are at all identical. Nations are as diverse as people. The conditions vary and so do the precise solutions, but beyond differences in history or culture, there is a clear similarity among agriculturalists everywhere. To others, there may seem to be little difference between the midwestern farmer working hundreds of acres with modern equipment and the grower with a small patch in a poor country, but if they met, they would recognize the ties. Their bond with the land, their attachment to the community, their goals to earn a decent living in agriculture, and their concern for their children's health, education, and future are the same. There are many reasons to support the effort of our world neighbors, for besides similarities, we have a common stake in an economically stable and more peaceful world.

When people are able to feed themselves, increase their incomes, and lead healthier lives, stability is the result. They are in a position to contribute to rather than drain from their country's progress toward development. As the U.S. and Europe (and nations like Brazil and South Korea) have shown, even population pressures decrease as citizens are able to sustain their families and become more productive.

It is no secret that **economic development** in other countries generates an almost immediate demand for products from this country. As people become more educated, they increase their chances of obtaining better-paying jobs, and a result is a demand for a higher quality diet and more food. Developing countries can become the basis for sound long-term markets for American goods. It is well documented that nations that in the past received and made good use of food, agricultural, and other development assistance have become our best customers.

The Importance of Developmental Assistance

Two countries which have received **developmental assistance** from America, Mexico and India, are now each able to purchase over one billion dollars of our agricultural products every year. Currently, 40 percent of our commercial farm exports go to developing countries, many of which can only afford to import due to recent economic

Figure 1–5 • Abdul Karim, a subsistence farmer in Bangladesh, and his sons put in the backbreaking work needed to grow rice. An irrigation project built with CARE's help has enabled Karim to increase crop production. *Source: CARE; photo by George Wirt.*

problems. Even with limited capacity, developing countries buy more U.S. exports than all of Western Europe. Through development, jobs are created and sustained both here and in the Third World, thereby contributing to our mutual prosperity.

Looking at one farm family in Asia, Ahmed Ali lives in eastern Bangladesh and typifies many of our farmer neighbors in the developing world; see Figure 1–5. Bangladesh, with its difficult colonial history, has evolved into one of Asia's poorest countries. By their independence in 1971, Bangladesh was teeming with refugees and new immigrants, largely because of the war for independence from West Pakistan.

Today, Bangladesh has 110 million people living in a land the size of Wisconsin. Whether poverty and over-crowding, or motivation and productivity, will prevail hinges on farmers like Ahmed Ali. Ahmed and his neighbors are determined to make a difference against difficult odds, and they have already shown what a farmer's resolve can do. With access to credit and basic agricultural education and extension, Ahmed Ali has learned "intensive cultivation methods." He has planted his seeds deeper and closer, and weeds them with extreme care. Last year, he and his family doubled their rice yield with just their own labor.

Ahmed Ali is not a lone initiater. Across the country other farm families are using hollowed-out bamboo for cheap irrigation piping. Piping which can last as long as five years in moist ground. This rural ingenuity and motivation mirrors our own community legacies. However, without assistance extended to Ahmed Ali in credit and agricultural services, his family's success and its benefits to the community might not have happened.

Across the globe in Guatemala, local farmers began growing a bean variety that produced unprecedented yields per acre. After careful testing, the variety showed such promise and durability that the government began recommending it nationally, along with seeds from foreign sources.

The record shows that farm families like these increased their productivity in the 1950s and 1960s, but suffered setbacks in the 1970s and 1980s due to the energy crisis, rising costs of credit, lowered earnings, and a diversion of government attention to the needs of industry. Add to that a drought in Africa far worse than our 1930's Dust Bowl, or a struggling cooperative in Asia with members who have little land and no capital, or a farmers association in Latin America with no access to agricultural services—and the hardships multiply.

A generation ago, President Kennedy said, "We have the manpower and technology to make world hunger obsolete, but we still lack the proven methods that can make the goal a reality." While methods of addressing food production and underdevelopment continue to be tested, many effective approaches have been discovered. The careful introduction of **appropriate technology,** an emphasis on the pivotal role of women, the opportunity for decentralized economic decision making, the sound use of food commodities in support of development, and a commitment to providing educational programs and training are all such examples.

The evolution of food aid serves as an example of our increasing understanding of the development process. Following World War II, the first massive aid campaign provided relief to a devastated Europe. Food and supply donations, like the well-known CARE package, helped that continent get back on its feet. Later, as many new nations of the Third World gained independence, American food aid was provided for the **Food for Peace Program.** This program, also known as Public Law 480, was legislated in 1954 under the Agricultural Trade and Development Assistance Act. Through it, millions of tons of nonfat dry milk, flour, butter, oil, beans, rice, and cornmeal were provided for relief efforts. American food assistance now includes wheat and other grain products as well.

Emergency food was and still is vital to situations such as Africa today where drought has brought death and starvation to hundreds of thousands of people. But beyond the immediate crisis, our food aid program is working in partnership with governments and local groups to support people's own efforts to lift themselves above poverty and to shape their own futures. Of utmost importance is aid utilizing food for the provision of meals for school children, which encourages attendance and promotes learning.

Today, women in many other countries play a full and active role in agriculture and agricultural development. Recognizing that women not only maintain their household but also work the fields, sell goods in local markets, and attend to livestock, educational training programs have been developed to assist in meeting their needs.

Figure 1–6 • Rural school children in Belize work on their garden, part of a new curriculum CARE helped develop that combines first-hand experience in farming, rabbit raising, and poultry production with academic subjects. One student remarked, "In my lessons I have found it easier to understand about soil and water, and to do sums about the gardens and chicks I know about." Belize is one of more than 35 countries throughout Asia, Africa, Latin America and the Middle East where CARE has a wide variety of feeding and development programs designed to help people survive and work toward self-support. CARE also provides disaster relief and primary health care. *Source: CARE; photo by Talibah Sun.*

In order for agricultural initiatives to reap positive results in the Third World economies, they must be supported by activities such as health care, education, and agricultural support through credit and marketing policies and advice. Local agricultural education and extension workers are excellent resources for sparking progress since they can provide hands-on help to obtain capital and analyze and solve problems; see Figure 1–6. Appropriate technology and financial services delivered through dedicated local workers can improve significantly the benefits of aid and assistance.

Conclusion

World agriculture needs to increasingly preoccupy international diplomatic, financial, and scientific circles. A new era is emerging. Knowledge, financial resources, and management capabilities are being mobilized nationally and internationally to achieve even greater breakthroughs in crop and animal productivity worldwide. It is important to understand that agriculture is fundamental to a nation's economic well-being and development. Every person, in the U.S. and abroad, is directly affected and influenced by agriculture or the lack thereof. We live and work in a complex and interdependent international environment.

The U.S. farm community understands what is required to create opportunities to satisfy basic needs and afford a better life. When the nation's farmers joined together, they created opportunities to obtain information, capital, and the means for mutual support. We are now in a position to share with others the experience and expertise gained over the years. This process of sharing can be an exchange of aspirations and ideas among agriculturalists around the world. The opportunities are infinite.

The transition from subsistence to commercial agriculture took Americans the greater part of a century to accomplish. Third World farmers must accomplish this in only a handful of years to succeed against the mounting odds of poverty. Support for agricultural solutions must acknowledge the motivation and hard work of local farmers to understand the problems they are facing, and to recognize that they will not work alone or in vain. Together with their resolve and our expertise and support, we can make the worldwide agricultural community a stronger force for global prosperity and peace in the years to come.

All of this will require vast investments in internationally coordinated education, research, and training—in addition to a commitment to preserving a healthy life-sustaining global environment. We must work together to make political, economic, and environmental decisions on the basis of science and rational thinking. In order to make the best decisions, we have to know the "what is" pertaining to global agriculture and learn from the "what was." Perhaps then we can better influence the "what should be" of global agriculture. According to Yuetter (1989), we have the communications tools, the knowledge base and the increasing levels of literacy to accomplish the needed tasks—if we reach out and help one another.

We will succeed if we work together to energize the tremendous reserves of human knowledge, applied skills, and optimism waiting to be called into action in the developing countries as well as in the industrialized nations.

Discussion Questions

1 Explain why the process of developing, transferring, and using agricultural technology requires knowledgeable people at all levels.
2 In other countries, why is the importance of agricultural education and extension so poorly understood and frequently given such a low priority?
3 Explain how the term agricultural education is generally defined and used by other countries.
4 Does the populace of the United States generally tend to be insular and inward looking? Why or why not, in your opinion?

5 Why is it no longer relevant that we consider ourselves to be a self-standing, independent economy among nations?

6 Do American farmers have little to fear and much to gain from agricultural improvement in other countries, especially the developing countries? Justify your response.

7 Concerning population growth, is there a sense of urgency that needs to be taken into consideration when planning agricultural development? Provide a rationale for your response.

8 Discuss the constraints with which American farmers had to contend during the 1800s and early 1900s. Explain the similarities of those constraints as compared to the constraints which farmers in developing and emerging countries are presently confronted.

9 Explain why economic development in other countries generates an almost immediate demand for products from this country.

10 Explain why we must work together with other countries to coordinate education, research, and training.

11 Explain the meaning of emerging nations.

CHAPTER 2
Population Growth and the Associated Food Deficit Problem

Source: United Nations Development Program; Photo by Ray Witten.

Terms to Know
overpopulation
agrarian society
demographic transition
family planning
metropolitan

Objectives

After reading this chapter, you should be able to:
- discuss the relationships between poverty and overpopulation.
- relate the reasons why parents in lesser developed countries want or need more children.
- describe the demographic transition which often occurs as nations overcome poverty.
- define the relationship between income and food consumption.

- identify the countries that will experience the highest population growth rates.
- discuss the impact that migration of rural people to urban areas has had within these urban areas in terms of overpopulation and unemployment.

The Population/Consumption Explosion: A Challenge to Global Agriculture

There are many factors that influence the demand for and consumption of food, including the level, growth, and demographics of population; and, the level and growth of incomes, tastes, and preferences deriving from different cultures. Consumption is a function of population, incomes, and policies that affects the prices and availability of food.

Future consumption patterns, and the consequent requirements on world agriculture, are dominated by the size and characteristics of the global population.

The global population is highly concentrated in a few regions. More than 85 percent of the world's people live in only 37 countries. Over three-fourths of those reside in the least developed parts of the world, and this concentration will most probably become even greater in the years ahead. Based on this prediction, 93 out of every 100 people added to the global population in the years ahead (from 1995 to 2025) will be in the least developed countries.

Even if demographers foresee the rate of population growth slowing slightly in the years ahead, the absolute increase in population will be larger than ever before. By the year 2000, the world will have to provide food for an additional 1.8 billion people. Figure 2–1 presents the rise in world population since 1950 and projects world population to be in excess of 6 billion people by the year 2000.

Overpopulation Defined

The population explosion can be identified as a major problem that all nations need to face in their struggle for global survival. It must also be recognized that a definite linkage exists with the consumption explosion of the richer, more highly developed industrialized nations. Each child born in the industrialized world consumes 20 to 40 times as much as a child born in the developing world. A small population increase in the industrialized world proportionally puts as much, if not more, pressure on world resources as a large population increase in the poor world.

Consequently, income often exerts a major influence on food consumption patterns. The relationship between income and consumption is fairly well defined. Where food consumption is already high, consumers eat very little more of most foods as their incomes increase. However, most of the world's people would increase consumption and improve their diets if they could. Many struggle just to meet basic caloric needs. But as their incomes grow, they begin to substitute meat, milk, and eggs for grain-based

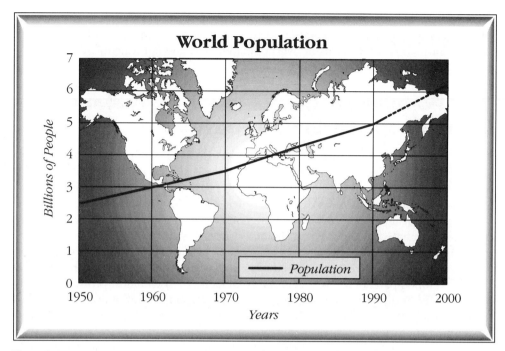

Figure 2–1 • World Population Graph. *Courtesy of Winrock International.*

foods. The level and distribution of income is thus most useful in helping to explain much of the sharply differing patterns of consumption and composition of foods among nations. The striking fact is that outside the industrialized countries, which have only a small proportion of the world's people but with high rates of consumption, most nations are struggling to increase economic growth, income, and food consumption.

Consumption will grow faster than production in most of the world's regions, and imports must expand to permit improved levels of living. Where the divergences are greatest, real food prices can be expected to rise. A more descriptive way of referring to the problem is to designate it the food-poverty-population problem. This can further be understood if we accept the definition of **overpopulation** as being that point when basic resources within its territory are insufficient to feed its people.

A basic aspect of overpopulation includes recognizing the fact that like all living things, people have an inherent tendency to multiply geometrically. In contrast, the supply of food rises more slowly, for unlike people, food does not increase in proportion to the existing rate of food production. This leads to the conclusion that the population is certain eventually to outgrow the food supply (and other needed resources), which may lead to famine and mass death unless a counter force intervenes to curtail population growth. One can argue about the details, but taken as a general summary of the population problem, the foregoing statement is one which no one can successfully dispute.

Nature of Population Growth in an Agrarian Society

The agrarian nature of many countries, such as the developing countries, generates pressures that automatically encourage large families. Mostly, these revolve around the basic role of children in the household. In **agrarian societies,** children represent productive assets. They provide low-cost labor services for the farm and also earn income from nonfarm employment where available. With high infant mortality, a large number of live births ensures an adequate number of living offspring. Furthermore, most developing countries do not provide care for the elderly. Therefore, a large number of children helps ensure old-age security for parents. To summarize, parents (particularly in developing countries) want more children in order to:

1. have, hold, and love.
2. satisfy parenthood tendencies.
3. have social security while impoverished—in the absence of social security, unemployment pay, sickness benefits, and old age pensions, poor people need children in order to look after them when they are out of work, sick, or old.
4. provide a human resource of workers.
5. avoid a mark of imperfection placed upon them, as small families are viewed in some countries.

Because education is not always rewarded or valued by some societies, parents may tend to discourage their children from attending school. Their rationale is based on the fact that time in school limits time for farm work and can also require money for out-of-pocket expenses for books and clothes. In other words, because developing countries require labor-intensive technology, children are needed for work.

Most older Americans who grew up during this nation's rural agrarian period will readily empathize with this type of value system. It was only after economic development occurred and the role of children changed that people in the United States (and other developed nations) found reason to voluntarily limit family size. The quality of life replaced subsistence as the family goal. In the process, children became more urbanized and fewer families and children worked the farm. To give children the education, training, and life-style thought appropriate, it became necessary to allocate family income more wisely. The high cost of such items meant the number of children must fall in line more closely with family resources if all were to benefit somewhat equally.

Population Growth Rates and Concurrent Food Deficits

As population growth rates in so many countries of the world continue at relatively high rates, the alarming concern must be where and how the food will be found to even sustain life, let alone to meet the goal of a more abundant life in a worldwide society.

Studies of present and projected world production of foods show that while production has been increasing, possible limitations upon the potential for continuing increases are such as to cause alarm and, in the judgement of some persons, to be of ominous portent. Compared to a rise in population of 1 percent per year in developed countries, the developing countries are increasing their population at about 2.5 percent per year. This phenomenal rise, coupled with a longer life span and a lowering of infant mortality as a result of modern medical technology, has largely offset slight gains in farm output.

An additional concern of many social scientists is the question of population quantity versus quality of life. It is generally accepted that apart from food resources, a high birth rate makes it more difficult to provide cultural opportunities and facilities for people. In other words, rapid population growth puts a strain on the social, economic, and cultural services of any society, and places an intolerable strain on developing societies. Social, economic, and cultural services are very important considerations to the achievement of quality human existence.

Factors and Policies That May Result in Lowering of Birth Rates

Cultural Attitude

Some studies of population growth seem to indicate that much depends upon the cultural attitude toward family size. If a large family is valued by one's peers and the opinion leaders of a community, that is an important factor in deciding the size of one's family. A small family may be regarded as a mark of imperfection, incompleteness, or inadequacy, By contrast, if the social pressures are in the opposite direction, a large family can be a source of embarrassment.

Demographic Transition

Because many researchers have concluded that a relationship exists between poverty and higher rates of population growth, more attention and increased study are being given to development as a major tool for achieving both hunger alleviation and an accompanying decrease in population pressures. This basic relationship between standard of living and rate of population growth is a major factor in the phenomenon known as the **demographic transition.**

To understand the way in which the process of demographic transition operates, certain associated factors should be understood:

1. reduced death rate increases the rate of population growth.
2. reduced death rate, particularly a reduced rate of infant mortality, may cause parents to opt for fewer children.

3. although a rising population increases demand on resources and results in worsening the population problem, it also stimulates economic activity. This in turn improves educational attainment levels, which again tends to increase the average marital age.

Some researchers see the effects of poverty alleviation and a consequent rise in educational levels and standard of living as a very definite and most desirable way of lowering a nation's birth rate. There has never been a population that, having acquired education and higher levels of living, did not reduce its birth rates. Unfortunately, the converse of this proposition is also true. We have yet to witness a population mired in illiteracy and poverty that has managed to reduce its fertility.

Utilization of Family Planning Services

Even though the premise is that eventually population growth control can be brought about by providing higher standards of living and eliminating abject poverty, the problem is too urgent and critical to depend upon this solution alone. Ample evidence is available to substantiate the claim that whenever **family planning** services are available to a population in which the use is clearly understood, a considerable portion of the people will utilize such services.

While wider availability of better family planning methods could rapidly reduce birth rates, additional efforts will be required to assure that people who are motivated to limit family size are also free to do so. Different approaches have worked for different cultures, but central to all of them is progress in reducing infant mortality. In addition, most successful approaches have also secured a greater role for women in family and community decision making by improving women's educational and economic opportunities.

Population Figures Shown

Frequently, a nation's population growth is analyzed in terms of percentage growth. For example, a nation may be experiencing a population growth of 2.9 percent annually whereas another nation may be experiencing less than 2.0 percent. Although the use of percentages is a tool utilized by demographers to project population growth, the presentation of actual numbers of inhabitants is much easier to comprehend. Also, percentages can be misleading. A densely populated nation that is experiencing a small percentage of population increase may in fact be adding more people yearly than a less densely populated nation that is experiencing a higher percentage of population growth. As you examine the table that follows, pay particular attention to the projected growth in population. You will notice the population of many nations will more than quadruple by the year 2100.

Table 2–1 lists the nations of the world by projected population in the year 2100,

Table 2–1 • Projected Populations for the Year 2100

Country	Year 2100	Year 1986
India	1,631.8	785.0
China	1,571.4	1,050.0
Nigeria	508.8	104.4
Soviet Union (Commonwealth of Independent States)	376.0	280.0
Indonesia	356.3	168.4
Pakistan	315.8	101.9
United States	308.7	241.0
Bangladesh	297.1	104.1
Brazil	293.3	143.3
Mexico	195.5	81.7
Ethiopia	173.3	43.9
Vietnam	168.1	62.0
Iran	163.8	46.6
Zaire	138.9	31.3
Japan	127.9	121.5
Philippines	125.1	58.1
Tanzania	119.6	22.4
Kenya	116.4	21.0
Burma	111.7	37.7
Egypt	110.5	50.5
Turkey	108.6	52.3
Algeria	104.0	22.8
South Africa	102.9	33.2
Thailand	98.9	52.8
Sudan	97.5	22.9
Uganda	79.5	15.2
Iraq	71.4	16.0
Germany	71.3	77.4
Afganistan	70.6	15.4
Nepal	70.1	17.4
Korea, South	69.9	43.3
Morocco	67.9	23.7
Mozambique	67.0	14.0

(continues)

Table 2–1 • (Continued)

Country	Year	
	2100	**1986**
France	62.6	55.4
Ghana	62.6	13.6
Colombia	59.3	30.0
United Kingdom	58.2	56.6
Italy	56.4	57.2
Saudi Arabia	54.9	11.5
Argentina	54.2	31.2
Malagasy Republic	52.2	10.3
Cameroon	50.3	10.0
Poland	50.0	37.5
Spain	49.1	38.8
Peru	48.0	20.2
Venezuela	46.1	17.8
Korea, North	45.5	20.5
Ivory Coast	45.4	10.5
Angola	41.4	8.2
Syria	41.1	10.5
Rwanda	38.5	6.5
Zimbabwe	38.3	9.0
Niger	38.1	6.7
Yemen, North	37.6	6.3
Malawi	36.2	7.3
Mali	34.9	7.9
Malaysia	32.7	15.8
Sri Lanka	32.4	16.6
Taiwan	32.3	19.6
Zambia	32.1	7.1
Canada	31.9	25.6
Romania	30.4	22.8
Burkina Faso	30.0	7.1
Yugoslavia	29.9	23.2
Somalia	29.5	7.8
Senegal	28.7	6.9
Guatemala	24.7	8.6

(continues)

Table 2–1 • (Continued)

	Year	
Country	**2100**	**1986**
Ecuador	24.6	9.6
Burundi	23.0	4.9
Guinea	22.8	6.2
Chile	21.4	12.3
Bolivia	21.1	6.4
Chad	21.1	5.2
Australia	21.0	15.8
Kampuchea	20.0	6.4
Benin	19.8	4.1
Czechoslovakia	19.4	15.5
Tunisia	18.4	7.2
Libya	18.1	3.9
Laos	16.9	3.7
Jordan	16.8	3.7
El Salvador	16.3	5.1
Sierra Leone	15.9	3.7
Netherlands	15.1	14.5
Dominican Republic	15.0	6.4
Togo	14.9	3.0
Honduras	14.8	4.6
Cuba	14.7	10.2
Haiti	13.9	5.9
Portugal	13.2	10.1
Central African Republic	11.7	2.7
Greece	11.6	10.0
Nicaragua	11.5	3.3
Hungary	11.0	10.6
Liberia	10.5	2.3
Bulgaria	10.1	9.0
Belgium	9.9	9.9
Congo	9.1	1.8
Papua New Guinea	9.0	3.4
Paraguay	8.0	4.1
Sweden	8.0	8.4

(continues)

Table 2–1 • (Continued)

Country	Year 2100	Year 1986
Israel	7.9	4.2
Mauritania	7.9	1.9
Austria	7.6	7.6
Hong Kong	7.4	5.7
Yemen, South	7.1	2.3
Albania	5.8	3.0
Lesotho	5.8	1.6
Ireland	5.7	3.6
Finland	5.6	4.9
Lebanon	5.6	2.7
Switzerland	5.5	6.5
Costa Rica	5.4	2.7
Mongolia	5.1	1.9
Puerto Rico	5.1	3.3
Denmark	4.9	5.1
Kuwait	4.9	1.8
Namibia	4.8	1.1
Jamaica	4.5	2.3
Norway	4.4	4.2
Botswana	4.3	1.1
Uruguay	4.3	3.0
Bhutan	4.2	1.4
New Zealand	4.2	3.3
Panama	4.2	2.2
Gabon	3.7	1.2
Oman	3.7	1.3
United Arab Emirates	3.6	1.4
Guinea–Bissau	3.4	0.9
Swaziland	3.4	0.7
Singapore	3.3	2.6
Gaza	3.2	0.5
Gambia	3.0	0.8
Djibouti	2.4	0.3
Trinidad and Tobago	2.4	1.2
Comoros	2.0	0.5

including their populations as currently estimated. Populations are given in millions of people.

Migration of Farmers to the Metropolitan Areas

Due to lack of profits or the inability to produce crops or livestock, many farmers and farm employees have been forced to leave rural areas in order to seek employment in **metropolitan** areas (The United States is no exception, see Table 2–2.). Furthermore, children who were reared in rural areas often seek employment in urban areas and also contribute to the aforementioned migration pattern.

Table 2–2 •

Largest Metropolitan Areas By Population
(Reported in Millions)*

Region	City	Population
North America	Mexico City, Mexico	19.3
	New York City, U.S.	18.1
	Los Angeles, U.S.	13.7
South America	Sao Paulo, Brazil	18.4
	Buenos Aires, Argentina	11.5
	Rio de Janeiro, Brazil	11.1
	Lima, Peru	6.5
Europe	London, England	10.5
	Moscow, Russia	9.3
	Paris, France	8.5
Asia	Tokyo, Japan	30.3
	Shanghai, China	12.5
	Calcutta, India	11.8
Africa	Cairo, Egypt	9.8
	Alexandria, Egypt	3.2
	Kinshasa, Zaire	2.9
Oceania	Sydney, Australia	3.7

*Source: National Geographic, 1991.

Mexico is a classic example. Approximately one-tenth of Mexico's land is suitable for agricultural production and a small fraction of that is suitable for permanent cultivation. Surprisingly, however, agricultural production in Mexico remains the mainstay of Mexico's economy. Large farms yield coffee, cotton, fruit, and vegetables. Mexico is also able to export much of the aforementioned crops.

Unfortunately, when crop prices fall and landholders, farmers, or farm employees cannot make a profit, they are forced to either sell their land or leave the farm and seek employment in the larger cities or urban areas. This migration of farmers contributes to the many problems of an already overpopulated area where unemployment is extremely high. Mexico City, for example, contains more than 19 million inhabitants in it's metropolitan area. It is projected that Mexico City will become one of the largest megalopolitan areas in the world.

Due to population booms and lack of jobs in nations like Mexico, people will seek employment (legally or illegally) in other countries. The primary concern of these refugees is survival and the need to feed their families. This is further evidenced by the large number of farmers who migrate north from South America and Latin America into Mexico and eventually into the United States.

The migration of farmers to urban areas will further contribute to the population/consumption problem for two reasons: (1) there will be fewer people involved in agricultural production and (2) there will be more people who will be without jobs.

By comparison, Japan (a highly industrialized nation) remarkably produces about three-fourths of the country's food even though less than one-fifth of the land is suitable for cultivation. Japan's population (approximately 123 million) is crowded into a space about the size of California. The largest metropolitan area in Japan is Tokyo (30.3 million people). The primary difference between Japan and Mexico is Japan's industrial capability has contributed to economic stability and employment opportunities. Compared to Mexico, Japan is an industrial giant that is capable of acquiring a sufficient amount of food for it's populace.

Conclusion*

Despite efforts to limit population growth, birth rates in developing nations continue at levels well above those that would permit stabilized population levels. Projections to the year 2000 and beyond are that population will continue to grow at nearly 2 percent a year in the developing world. Additionally, significant reductions in population growth will be difficult to achieve in these nations for a couple of reasons.

One person's family size does not create a population problem in Bangladesh. It is the collective impact of individual actions within the group that creates the excess. It is difficult to get one individual to see this and to act in the group's interest. In the

---

*Adapted from the National Association of State Universities and Land Grant Colleges.

developing nations, limiting family size may not be in the best interest of individual families, particularly in rural areas. In most societies, the decision to have children is a free choice, unencumbered by public relations. Only in a few nations such as China is this decision influenced sufficiently by government and public policy to have a meaningful impact on population growth.

If incomes and education for rural residents can be improved, value systems can change and voluntary control of population can occur. Strong evidence worldwide shows that family planning and voluntary control of family size increase dramatically with improved education and higher incomes. Until these advancements in the lives of people become fairly widespread, it appears the population trends of the past 30 years will continue. Efforts at control through family planning will have some impact. But the problem is enormous. The poorer areas of the world will be forced to deal with the desperate problems of how to feed a rising population.

Discussion Questions

1. Compare rates of population growth in (a) developing nations and (b) developed nations. Which countries are experiencing the fastest population growth?
2. Why is the population growth problem more correctly referred to as the food-poverty-population problem?
3. How would you define overpopulation? Is the United States overpopulated according to your definition?
4. How does food production trends compare with population trends?
5. Discuss why parents in developing nations want or need many children.
6. What is meant by demographic transition?
7. Explain why the encouragement of public school attendance may bring about a decrease in population growth.
8. What is meant by cultural attitude?
9. How do income changes around the world influence demand for individual agricultural commodities?
10. Explain the relationship between population explosion and consumption explosion.
11. What impact does the migration of farmers and farm families have on metropolitan areas?

CHAPTER 3
The Worldwide Challenge of Providing for Adequate Human Nutrition

Source: Dekalb Plant Genetics.

Objectives

After reading this chapter, you should be able to:

- list the common manifestations of the imbalance between the supply of food and the number of people.
- identify and describe the causes of world hunger.
- describe the effects of malnutrition upon human beings.
- identify the areas in the world having chronic food shortages.
- describe each phase of the hunger cycle and its impact upon the mental and physical well-being of individuals affected.
- discuss steps or actions that nations might take in order to overcome malnutrition.

Terms to Know

world food problem	maternal depletion	illiteracy
world hunger	syndrome	culture
infant mortality	hunger cycle	food
malnutrition	poverty	idle land

Concept of a Global Responsibility

Our understanding of the world is changing. We are connected to global events regardless of where or when they happen.

We have been brought closer to the realities of agriculture in many parts of the world by vivid pictures of the famine that struck parts of Africa in the 1980s. We have seen the interruption of food production cycles caused by drought, flood, and hurricane. The resulting human devastation has been enormous. While once brought to our attention in word and pictures in print media, we are now confronted by the sights and sounds of natural disaster as it happens via satellite communication.

What we are gaining is a new view of the world. Far away places are not so far away. Strange customs are not so strange. Foreign faces and foreign speech do not seem as foreign. We are beginning to see how it all fits together. (Adapted from Edgar Persons, 1990.)

Granted, more poeple—and a higher proportion of the world's population—enjoy adequate nutrition in the 1990s than ever before in the world's history (Avery, 1991). Indeed, the entire decade of the 1980s saw record crops around the world, driven by a host of new plant breeding advances, widespread use of improved farming systems, seed farms, and roads. World food production has doubled since 1960, thanks largely to advances in food production technology (Avery, 1991). Though there has been a 26 percent gain in per capita food supplies in the developing nations, there remains much to be done relative to world food production, trade, and development. We must analyze the problem before attempting to offer a solution. We have to know what it is that we are a trying to accomplish agriculturally, and for what reasons.

The concepts of one nation helping another and of a global responsibility are fairly recent phenomena that have evolved with the development of a highly interrelated global food system (Wennergren and Furlong, 1985). This evolution resulted from nations being linked by sophisticated communications, international financial and commodity markets, efficient and low-cost international transport systems, agricultural production systems, and highly skilled scientific institutions. The development of this global food system, rather than resolving the issues of the world's food supply, has made each nation more visible to us and created a greater realization of our interdependence.

The World Food Problem Defined

The **world food problem** is defined as a shortage of food that threatens the welfare of large segments of the world's population. Starvation, hunger, and malnutrition are common manifestations of the imbalance between the supply of food and the number of people. It has been estimated that as many as 500 million people suffer from hunger and the effects of starvation. The World Bank estimates that probably 800 million people in the developing world alone live in absolute poverty. For the most part, these conditions are chronic and are imposed on these people as a daily way of life.

All nations have poor and hungry segments in their populations, but these are not generally identified as trouble spots when discussing the issues of the world food problem.

The demand for food is an economic concept that recognizes that people have many different needs. These needs must be satisfied by limited purchasing power (Wennergren and Furlong, 1985). For individuals, the demand for food depends on both their ability and willingness to pay for food. This becomes complicated because the combined effects of changes in population and per capita income determine the magnitude of the demand for food that must be met worldwide. This demand for food is reason enough for agriculturalists to study the continued need for food as well as learn more about the people who are demanding it.

World Hunger Defined

Several times in the last 30 years, the world has experienced intense concern about relative food scarcity and troublesome surpluses. The highly uneven distribution of food around the globe remains a persistent concern. Today, we enjoy an unusual agricultural abundance in the food surplus regions of the world, but the potential is perhaps greater than ever for a sudden dramatic shift. Global food stocks provide little protection for many of the world's people, and that shield is diminishing as populations continue to grow.

It is extremely important to monitor the global food situation closely. In particular, we must better understand the underlying trends determining the ever-precarious food balance, which can shift so quickly and dramatically.

Those countries that have experienced food scarcity have been confronted with the ultimate challenge of providing for their populace.

A report of the United States Presidential Commission on World Hunger (1980, p. 26) suggested the following as a response to the question, "What is world hunger?" It is not the kind of hunger you feel when you miss lunch. Many, perhaps most, of the nearly quarter of the human race suffering from malnutrition and starvation feel no pangs of hunger at all. They are often physically and mentally too far gone.

But, even among the poorest of the poor, there remains a fraction with the energy and leadership needed to protest their situation in the political arena (where permitted) or with violence. Whether the hungry ones are passive or aggressive, whatever their race or location, they share a life condition known as **world hunger** which has been defined by the World Bank as characterized by malnutrition, illiteracy, disease, squalor, high **infant mortality,** and low life expectancy beneath any reasonable definition of human decency. The U.S. National Academy of Sciences asserts that sufficient supplies exist to feed everyone if they were equitably distributed.

Other current indicators of world hunger are equally discouraging. In some poor countries, 40 percent of the children die before they are five years old. For those who survive, their average life expectancy is 48 years.

Meanwhile, the richer nations consume 50 percent of the world's food, although they account for only 30 percent of the population. The average person in the poorest countries produces $150 worth of wealth in a year while the richest 20 percent of the population typically control half of the national income.

Food deficits in the Third World are expected to get worse as these populations continue to grow. Of particular concern are the African countries (Table 3–1). Africa has the highest yearly rate of natural population growth of any developing region, averaging 2.9 percent. That growth is expected to continue because the factors encouraging high fertility will change very little. Africa is also the only region of the world where per capita food production has declined over the past 20 years.

In addition to thinking about hunger in terms of a given country or region, it may be helpful to think in terms of people as individuals. The U.S. Presidential Commission Report (1980, p. 2) provided this very succinct picture:

In the next 60 seconds 234 babies will be born

- 136 in Asia
- 41 in Africa
- 23 in Latin America
- 34 in the rest of the world

23 of these 234 will die before age 1

- 6 in Africa vs. 0.01 in North America
- 2 in Latin America vs. 0.025 in Europe

34 more will die before age 15. 50 to 75 percent of these deaths can be attributed to a combination of malnutrition and infectious diseases.

Many who do survive beyond age 15 will be stunted in growth and will suffer brain damage that can incapacitate them for life.

Malnutrition Defined*

The primary world health problem today is **malnutrition.** Malnutrition at its fundamental biological level is the inadequate supply of nutrients to the cell. A lack of essential

---

*Adapted from Sue Rodwell Williams book *Nutrition and Diet Therapy*

Table 3–1 •

Nations Most Seriously Affected by Hunger and Malnutrition

Africa:

Central African Republic	Kenya
Chad	Lesotho
Benin	Malagasy Republic
Ethiopia	Mali
Ghana	Mauritania
Ivory Coast	Niger
Rwanda	Senegal
Sierra Leone	Sudan
Somalia Democratic Republic	Tanzania
United Republic of Cameroon	Burkina Faso

Latin America:

El Salvador	Guyana
Haiti	Honduras

The Far East:

Bangladesh	Cambodia
India	Laos
Pakistan	Sri Lanka

The Near East:

Yemen

---

nutrients at the cellular level is the result of a complex web of factors: psychological, personal, social, cultural, economic, political, and educational. Each of these factors is an important cause of malnutrition. If these variables are only temporarily adverse, the malnutrition may be acute, and alleviated rapidly, leaving no long-standing results or harm to life. But if these variables are continuously adverse and unrelieved, malnutrition becomes chronic. Irreparable harm to life follows, and eventually death ensues.

Human misery and death caused by malnutrition are more stark in some regions than in others; however, malnutrition occurs everywhere. These effects are more widespread in less developed areas of the world, but are present in the more developed nations also.

The fundamental cause of malnutrition is lack of food. Certain nutrients in food that are essential to the sustenance of cellular activity are missing. Various factors may cause or modify this lack of food.

Food quantity. The total level of food ingested may be below the level required to maintain body tissues. The food deficiency may be partial or complete, seasonal or constant.

Imbalance between community food supply and need. The amount of food available per person may be reduced by natural disaster (drought, flood), or by man-made disaster (war, overpopulation, poverty).

Food quality. The food available may not be fresh or may have low nutrient value.

Food timing. The food may not be presented when needed, such as for infant and child feeding, or may not be in proper balance.

Malnutrition results not only from sustained deprivation of necessary calorie and protein levels, but also from dire scarcity of required vitamins and minerals.

Two separate problems seem to emerge from nutritional studies of the chronically underfed:

1. infant malnutrition with subsequent inadequate growth and development.
2. nutritional imbalance in the diets of adults.

Obviously, there are many degrees of malnutrition, ranging from mild to fatal. Healthy people can tolerate the loss of about one-quarter of their weight, but when they lose more than this, they become ill and their life may be threatened.

Severe malnutrition of an infant occurring either in the prenatal period or in the first few months of life has been definitely established as a causative factor in abnormal development of the brain and nervous system. Malnutrition in animals and children permanently retards brain growth and results in abnormalities in later development. It now seems that there is a critical period during early development when the child is most susceptible to such environmental insults. During this period, poverty and its accompanying evils (hunger, sickness, and lack of emotional stimulation) permanently stamp a child and may cause damage from which there is no recovery.

The poor remain poor partly because during their critical development period, malnutrition, poor health, and the components of poverty have cheated them of the chance to fully develop their potential and extricate themselves from the cycle.

Maternal Depletion Syndrome

When scant supplies of nourishing foods for mothers are combined with oft-repeated pregnancies, a condition known as the **maternal depletion syndrome** frequently occurs. Closely associated with this condition is the tragedy of the deposed child.

Assuming that a woman's babies are born alive and that she herself survives, repeated pregnancies coupled with prolonged lactation put a drain on a woman's proteins, calcium, and iron—the maternal depletion syndrome. This results in damage to her babies by her own premature aging, and often in her death as a still relatively young woman. This syndrome is seen in all social classes and is most marked in younger mothers. Children from large families are smaller and shorter than those born into small families; the addition of each new child adversely affects the development of his siblings.

The Hunger Cycle

Let us examine a vicious cycle, the **hunger cycle.** As portrayed graphically by United Nations documents (refer to Figure 3–1), one baby in six is born underweight and is

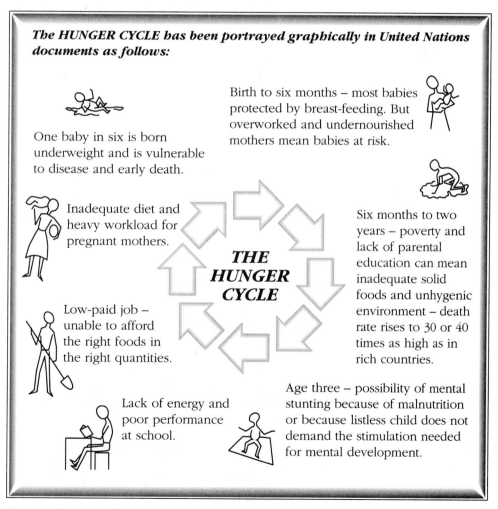

Figure 3–1 • The Hunger Cycle. *Source: The United Nations Development Program.*

vulnerable to disease and early death. From birth to six months of age, most of the babies are protected by breast-feeding; however, because the mothers are generally overworked and undernourished, the babies are even at greater risk. From six months to two years of age, the children may receive an inadequate amount of solid foods and may live in an unhygenic environment due to poverty and lack of parental education. By the time the child reaches three years of age, mental stunting may result due to malnutrition. The lack of adequate nutrition is the primary reason for the child's lack of energy and poor performance when they begin school. Unable to perform well in school, the child grows to become an adult who has to settle for low-paying jobs and therefore unable to afford enough food. Finally, pregnant mothers who endure inadequate diets and heavy workloads may give birth to a baby who is underweight and vulnerable to disease and possible early death. At that point, the hunger cycle begins again.

Broader Causes of Worldwide Hunger and Malnutrition

Poverty

Almost all knowledgeable scholars and researchers of world hunger agree that **poverty** is the most potent factor associated with problems of malnutrition. These scholars and researchers also agree that the primary cause of world hunger is poverty. Even in years of abundant harvests, many of the world's hungry have neither the land on which to grow food nor the money with which to buy it; see Figure 3–2. In many rural areas, large numbers of people often have little or no access to land, water, or credit. In urban areas, where schools, medical care, and food may be available, jobs (and thus the income to obtain these services) are scarce, largely because so many men, women, and families have been forced to leave the countryside. Inequitable distribution of resources—particularly land—is a major cause of poverty.

Illiteracy

In contrast to most nations of the developed world, especially the United States where agricultural education and extension have achieved a high level of knowledge dissemination, the educational portion of agricultural education and extension in the majority of developing nations is disappointingly meager. Effecting change in production practices which enhance higher agricultural productivity is a challenge. Undoubtedly, one factor which hampers the effectiveness of delivery systems in developing nations is the incredibly low level of both general and specialized education. **Illiteracy** is a major factor that influences food production potentials in many of the developing nations. In recognizing this fact, Mexico spends one-fourth of its federal budget on education. In contrast, some of the developing countries allocate less than one dollar per year to educate each child of elementary school age.

Figure 3–2 • Like many African women, these women in Sierra Leone juggle their responsibilities as wives, mothers, and farmers, often keeping an eye on their children as they work in the fields. Some 80 percent of the food produced in Africa is grown by women. The average farmer in Sierra Leone earns less than $100 a year, working on a landholding of less than five acres. *Source: CARE; photo by Vera Viditz-Ward.*

Recently, the U.S. Agency for International Development reported that in spite of the fact that education is one the largest single items of public expenditure in the developing countries, more than half of the population in most developing nations have never been to school; illiteracy in some countries of Asia, Africa, Latin America, and the Middle East is increasing because school enrollment cannot keep pace with high population growth; and even where schooling is available, the conditions and quality of learning are so poor that up to 50 percent of the children enrolled in school fail to complete their third year.

Cultural Forces*

A century old definition describes **culture** as that complex whole that involves knowledge, belief, art, law, morals, custom, and any other capabilities and habits acquired by humans as members of society. Modern anthropologists have enlarged the definition of culture to include the entire way of life of a people. Often the most significant thing that can be known about culture is what it takes for granted in daily life.

These many facets of a people's culture are learned, and are developed over a long time period. Culture is partly the result of the environment. The environment may be harsh and hostile, and the way of life developed is what enabled people to survive. Sometimes the changing of these habits by an outsider who does not understand these adaptations may upset this balance with nature. This has been the case when people have tried to impose Western culture and habits upon people in other parts of the world, without prior study and appreciation of established customs. Such programs have failed for that reason.

Food in a Culture. Food habits are among the oldest and most entrenched aspects of many cultures. They exert deep influence on the behavior of the people. The cultural background determines what shall be eaten, as well as when and how it shall be eaten. Beneficial and injurious customs are found in every part of the world. By and large, food habits are based upon food availability, economics, or symbolism. Included among these factors are the geography of the land, the agricultural practices, the economy and market practices, and the history and traditions.

Cultural Determination of What Food Is. Items considered to be **food** in one culture may be regarded with disgust, or may cause illness in the persons of another culture. In America, milk is valued as a basic food; in many other cultures it is rejected with revulsion as an animal mucous discharge. In the Philippines, the Ifugao tribesmen of northern Luzon are known for their enjoyment of dietary items they prize, such as dragon flies and locusts, which they boil, dry, and grind into a powder. They also fry crickets, flying and red ants, beetles, and water bugs even though they are proficient in the production of rice.

---

*Adapted from Sue Rodwell Williams book *Nutrition and Diet Therapy*

Religious aspects of culture also control food rejection and use. Pork, for example, is unacceptable as food for a Moslem or an orthodox Jew and any meat is unacceptable for the Seventh Day Adventist. The strict Hindu or Buddist eats no meat, and even the liberal Hindu may not eat beef.

The Slow Pace of Appropriate Technology Transfer and Development

One of the most striking differences between farming operations in the developed and developing nations is the technology used. There is a stark contrast between the large labor-saving machines on farms in the U.S. and the hand tools or oxen-powered equipment of many developing nations.

Governments in developed countries, in cooperation with international organizations, have helped to improve agricultural production in developing countries by sharing their biological and technological advances; see Figure 3–3. But the wealthier countries have become aware that technology that works for them may not work everywhere. A large sophisticated tractor that is ideal for a sprawling Iowa cornfield would be useless and unserviceable on the tiny farms in the more remote areas of the world where labor rather than capital is abundant.

The key is to develop and transfer appropriate levels of technology to other countries. The kind of machinery must fit their farm size, be easily operated, and be serviceable by local personnel with locally produced parts; see Figures 3–4 and 3–5.

Steps Nations Must Take to Overcome Malnutrition and Reduce Their Need for Imported Food

Overcoming the ravages of malnutrition is a complex problem for any nation. Each of the steps presented here are discussed at length in later parts of this text.

First, all nations must give high priority to agricultural research, education, and production. A large number of the food problems of the world are the result of governmental decisions rather than immutable forces. Sufficient resources and technology presently exist to increase food production substantially in all nations. The major problem is to develop food production in those countries which are faced with the most serious food shortage (Table 3–2).

Each nation should remove obstacles to food production and provide proper incentives to agricultural producers. To increase food production may call for land reform measures, tax credit changes, and more educational opportunities in agricultural development.

Second, develop **idle land** for agricultural purposes. Idle land refers to productive land not currently being utilized for production purposes. Much idle land is available to

Figure 3–3 ● The Food and Agriculture Organization of the United Nations (FAO) helped Mexico to develop a pilot plan for improving the management of valuable cedar and mahogany resources in Quintana Roo, one of the least developed states of Southern Mexico. FAO provided consultants through its Technical Cooperation Program (TCP) to help establish tropical nurseries, plantations, industries, and training in tropical logging methods. The plan evaluated local agri-forestry practices, sought to integrate forest land use with public lands, and helped adapt the national forest management program to tropical zones and to the needs of local farming communities. The plan also seeks to conserve natural fauna, reduce soil erosion, reduce rural population migration by creating community industries such as reforestation projects and coffee cultivation, reduce destructive "slash-and-burn agriculture," and to introduce alternative energy resources, including biogas production. *Source: FAO; photo by L. Taylor.*

be developed for profitable crop production. Such land can be made productive by clearing undesirable growth, leveling, irrigating, and fertilizing.

Third, increase food crop yields. The yields of food crops in most of the poor countries are extremely low. The greatest opportunity for increased food production is to increase the per-acre yield of the land already in production.

Fourth, increased emphasis must be given to family planning and population education.

Fifth, the status of farmers must be raised. The socioeconomic rank of farmers must be improved. This may be accomplished by making farming more financially rewarding,

Figure 3–4 • The FAO has helped the Ivory Coast to implement a market gardening development plan. The plan's aim is to increase crop production to eventually meet growing population demands, promote and modernize the farming community, and to lessen the disparity of vegetable and fruit production between regions. A farmer adjusts the irrigation sprinkler system for a large market vegetable garden. *Source: FAO; photo by F. Mattioli.*

giving special recognition to outstanding farmers, and conducting tours for local farmers to observe successful farming operations.

Sixth, we all need to understand that today's dynamic and increasingly complex world food economy is central to many of the most critical concerns of individual nations and the international community at large. Examples abound:

- A severe drought across broad areas of the African continent threatens the already-meager food supplies of large numbers of people.
- Western Europe and the United States continue to escalate their confrontation over agricultural trade matters.
- The U.S. and the Commonwealth of Independent States (formerly the U.S.S.R.) negotiate new long-term trade agreement for agricultural products, an action with broad significance for continued relations.

Figure 3–5 • Introduction of Jojoba in the Republic of Yemen in 1990. This intercountry project, which was started in 1985, is financed by the United Nations Development Program (UNDP) and executed by the FAO. Its aims are the development of marginal lands, soil conservation, development of agri-industry and generation of additional income for the rural population in semi-arid regions through introduction of new crops such as Jojoba. Pictured is a Yemeni farmer examining an outstanding growth of Jojoba in his two-year-old plantation, which was planted by the FAO Regional Project for introduction of Jojoba in the Arab region. *Source: FAO; photo by Hussain S. Ahmed.*

Table 3–2 •

Production/Consumption Comparisons of Selected Countries*

Country	Percent of Population in Agriculture	Percent of Total Personal Spending	Minutes Work to Earn		
			1 lb. Pork	1 lb. Bread	1 dz. Eggs
United States	2.0	13.3	12	4	4
China	71.0	60.0	180	NA	200
Italy	11.0	30.3	20	6	11
France	9.0	19.2	16	8	13
Canada	5.0	14.5	8	3	6
Mexico	37.0	40.9	36	7	16
Australia	6.0	16.6	19	4	13
Brazil	40.0	41.0	54	14	19
Britain	2.0	17.9	25	5	15
Soviet Union (the Commonwealth of Independent States)	15.0	31.0	NA	8	71
Japan	13.0	21.5	41	40	11

*Note: Statistical comparisons of this type remain an inexact science because we are comparing countries with cultural, economic, and religious differences, all of which significantly influence the production, distribution, and consumption of food products. Source: Agricultural Council of America.

- The U.S. government has instituted the largest agricultural crop production control program in history.
- The economic situation in many developing countries becomes more critical, especially for those dependent upon external markets for basic raw commodities and those that must import petroleum in ever greater quantities as a prerequisite for future development.
- The U.S. agribusiness community continues to ponder investment decisions and strategies in order to enhance the export of agricultural products.

The global food economy is dynamic, always undergoing substantial and complex changes that will affect performance in the years ahead. It will remain instrumental in nutrition and health concerns, economic stability, and even the political stability of much of the world.

While no one can make precise predictions about the future course of so dynamic and complex a system, it is possible to examine fundamental forces that drive the system and to develop broad outlines of how it may evolve in the years to come. Having some notion of the future world environment is important to most endeavors, whether they are constructing forward-looking national policies, developing marketing and investment strategies, or assisting in improving the lot of the world's people.

Conclusion

There is a compelling need to provide adequate nutritious food for all people of the world. Many national and international agencies are working toward this objective. All governments must accept responsibility for hunger and malnutrition, which affect many millions of people. This should be an objective of the international community.

Discussion Questions

1. Define malnutrition.
2. Describe the relationship between hunger and poverty.
3. Describe the vicious cycle often maintained in Third World countries brought about through the poverty-malnutrition syndrome.
4. What nations and areas of the world are most seriously affected by insufficient food resources?
5. Explain some of the broader causes of worldwide hunger and malnutrition. Explain why it is important for agricultural students (in the developed countries) to understand these causes.
6. Discuss some of the steps or actions that developing and developed nations might take to overcome constraints upon hunger alleviation.
7. What is meant by the maternal depletion syndrome?
8. Why is the world food economy so central to many of the most critical concerns of individual nations and the international community at large?

CHAPTER 4
The Nature and Extent of Agricultural Production Worldwide

Source: Food and Agriculture Organization; Photo by L. Callerholm.

Objectives

After reading this chapter, you should be able to:

- compare and contrast selected farm types, sizes, and operations in other nations.
- explain why farmers in other countries do not produce food in the same manner as in the U.S.
- differentiate among the various types of farming systems utilized throughout the world.
- discuss the extent and distribution of crops and animals providing the world's food.
- explain the interrelated factors that affect the worldwide production of food.

Terms to Know

Dutch Polders
industrial crops
hectares
infrastructure

subsistence farming systems
commercial farming systems

collective farming systems
cereal grains
aquaculture

Introduction

A newly constructed hut in Papua, New Guinea is looked upon by an out-of-country visitor with favor as he contemplates the pride of the native family which soon will be occupying it. But closer inspection reveals the hut already has hundreds of squiriming occupants feasting greedily upon mulberry leaves. Through an interpreter, a caretaker-operator carefully relates how important certain techniques are to silkworm production; see Figure 4-7 (page 76).

In Zimbabwe, a visitor may view a small cattle herd relishing the forage found on communally-owned grazing lands. Although the small herds are individually owned, their composition is characterized by a high percentage of oxen; this is due to the fact that their farming system is based upon traditional labor intensive technology with limited capital input.

At another locale in west central Kenya, the visitor will find laborers meticulously picking leaves from small trees or bushes in order that far away gourmet tea sippers may find satisfaction and contentment.

In Holland, intensively operated farms (on small acreages) will provide a riot of color as workers apply special techniques to produce flower bulbs to meet the demands of flower fanciers in other parts of the world.

For a fifth glimpse of an agricultural endeavor, workers in Bangladesh produce jute, which requires them to work in the channel vats stirring and rearranging the rotting stalks; not a very pleasant task!

Only one of the agricultural enterprises featured can be classified as a food crop, and only indirectly at that. When the range of enterprises encompassing agricultural production is considered, categorization may be shown as follows:

1. food crops, edible animals, fish, and poultry.
2. forestry and fiber crops.
3. non-nutritive substances, articles, and animals for usage other than food.

World agriculture is a complex, many faceted phenomenon. Inclusive of the various phases of production, processing, marketing, and distribution, intricate relationships exist. The magnitude of diversity characterizing operations in the global agricultural industry becomes apparent. In this chapter, the nature of such diversity is presented featuring a few selected areas.

A Global Comparison of Selected Farm Types, Sizes, and Operation*

Because many U.S. students select industrialized nations to participate in educational or work exchange programs, a comparison of selected farm types, sizes, and operations

---
*Adapted from Communicating for Agriculture, Inc., 1991.

among these countries will be featured. A few examples from developing countries have been selected also for comparison.

An Introduction to Europe

Like the U.S., farm types and sizes vary a great deal within and between European countries, but most farms are from 30 to 500 acres in size, with the smallest in Norway and Switzerland, and the largest in Sweden and France.

Generally, herd sizes vary from 15 to 200 head of cattle, 20 to 500 head of sows, 100 to 2,000 feeder pigs, and 100 to 500 sheep. Weight gains and feed efficiency is often high in European countries due to the European farmers' resource consciousness and individual attention to each animal.

Crop yields are also quite high due to intensive fertilization, plant protection programs, abundant rain, and irrigation.

Most livestock, including cattle and sheep, are kept indoors during the winter months. This means a considerable amount of manual work with milking, feeding, barn cleaning, and individual attention at calving and farrowing.

The larger livestock operations have a high level of indoor mechanization, including feed mixers, barn cleaners, and mechanized feeding systems. Practically all dairy farms have milking machines, although most are based on bucket or pipeline systems.

In most countries, pigs on a large farm are kept in buildings with slatted floors and slurry handling systems. On modern farms, farm feed mixing, mechanized feeding systems, and efficient insulation and ventilation are also common.

Field mechanization is based on 25- to 125-horsepower machinery and two- and four-wheel drive tractors, depending on the size and type of farming. A great variety of implements are generally available to meet the requirements of all crops. Complicated on-farm repairs of tractors and machinery may not be as common in European countries as they are in the U.S. because most repair facilities are usually within reach and the smaller farms can not justify extensive repair facilities on the farm.

In brief, European agricultural endeavors are generally as modern, sophisticated, and complex as agricultural endeavors here in the U.S. They too have a totally integrated food and fiber system. Let us examine production agriculture more closely in some of these European countries.

Denmark. Denmark is a small country consisting of a main peninsula and nearly 500 islands, of which 60 are populated. The Danish countryside is either quite flat or gently rolling. More than 70 percent of the country is highly productive farm land, only occasionally interrupted by forests, lakes, and towns.

Farmland is considered an important national asset in Denmark. Before young farmers can purchase land, they must earn a farming license. Also, Danish farms have long specialized in exports.

The typical farm is in the 110- to 500-acre range, specializing in four to six different crops and one type of livestock, normally dual-purpose dairy cattle or pigs. In spite of its size, Denmark is the world's largest exporter of bacon and the fourth largest exporter of dairy products.

Danish crop production is dominated by winter and spring varieties of barley, which is grown on about 60 percent of the farm land. On the remaining 40 percent, a great variety of crops are grown including wheat, rye, oats, fodder beets, sugar beets, canola seed, potatoes, cultivated grasses and legumes, silage corn, and a wide range of seed crops.

Due to intensive fertilizer use and the relatively high rainfall, crop productivity is high; however, during dry periods irrigation machines are extensively used, especially on the lighter soils.

Farm mechanization is mainly based on imported 50- to 125-horsepower tractors, Danish and imported combine harvesters, and Danish-made implements.

The climate is practically the same all over the country—fairly mild winters with some frost and snow, and relatively cool summers with many rainy days.

Norway. The country is mostly mountainous with lots of forests and breath-taking scenery. Only about 3 percent of the country is suitable for agricultural purposes. Most farms, therefore, are located in narrow valleys and in the plateaus of southern Norway.

The typical farm size is in the 20- to 90-acre range, specializing in two to five crops, some permanent grazing, and one or two types of livestock, normally dual-purpose dairy cattle, pigs, poultry, sheep, or goats.

Common crops include barley, oats, potatoes, fodder beets and green crops for direct feeding, silage, or hay. Norwegian farmers probably lead the world in anhydrous ammonia treatment of straw to increase the feed value to its maximum. In addition, large quantities of concentrates and roughages are imported.

Farm mechanization is based on 20- to 90-horsepower tractors adapted for forestry work as well as agriculture, and many are equipped with four-wheel drive. Most farm implements are made in Norway or imported from Denmark.

Many farm families combine farming with forestry and are proficient at thinning forestry stands, felling trees, and hauling timber to the roadside. Most forestry work is done with tractor-adapted tree handling equipment and powerful chain saws.

The climate varies greatly with altitude and latitude, but weather conditions in most agricultural areas are temperate due to the effects of the Gulf stream in the Atlantic Ocean. Annual precipitation is fairly high, and the ground can be snow-covered up to five months of the year.

Norway is not self-sufficient as far as food is concerned. Its official policy is to maintain rural population and agricultural production at the highest level. This is achieved by offering extensive subsidies at production and marketing levels to secure a farm family equal to that of industrial workers. This explains how a farm family can maintain a good standard of living on 20 to 30 acres with seven to eight cows. A government-supported farm relief service also enables farmers to take summer vacations.

Today, Norwegians live in one of the world's richest countries. The government provides free education and health care, and unemployment remains low.

Sweden. Sweden has great variations in climate and topography. The countryside is generally flat or undulating, but quite beautiful with thousands of lakes and extensive forests.

Although less than 8 percent of Sweden's total land surface is suitable for cultivation (and most of this is found in the southern third of the country), its farms furnish nearly all food required.

The typical farm is in the 110- to 500-acre range, specializing in three to five crops and one to two types of livestock, normally dual-purpose dairy and beef cattle or pigs.

Common crops include winter and spring varieties of wheat or barley, oats, canola seed, seed crops, sugar beets, and various forage crops for green feed, silage, or hay. Most protein supplements are imported.

Farm mechanization is based on 50- to 120-horsepower tractors, many of which are made in Sweden, and a high percentage of implements and combine harvesters imported from Denmark or Finland.

The climate is fairly uniform. Summers are relatively cool and rainy, and winters are reasonably mild with snow falling three to four months of the year. Some winter work includes clearing snow from roads and farmyards.

Sweden is nearly self-sufficient in most food products and has a small surplus in some. The government protects domestic production and subsidizes consumer prices to assure farmers have an income comparable to that of other groups of society.

Netherlands. Netherlands means "lowlands," and true to its name nearly half of this small, flat country lies below sea level. Over the centuries, the Dutch have learned how to protect their land and how to reclaim more from the sea.

The Dutch countryside is extremely flat and largely shaped through reclamation of swamps and coastal areas. Water is drained into the ocean through an extensive canal and pumping system. About 50 percent of the highly productive farm land in Netherlands is located below high-tide level; however, thanks to the Delta Plan, an ambitious flood control project, these areas are safe behind the dikes.

Nearly 70 percent of Netherland's total area is utilized for agricultural purposes. The typical farm is 75 to 165 acres, specializing in four to six crops and one to two types of livestock, with dual-purpose holstein dairy cattle being the most popular. Other livestock includes beef cattle, pigs, sheep, and poultry.

Common crops include wheat, barley, oats, rye, potatoes, sugar beets, flax, grass and forage crops for green feed, silage or hay, corn, and various seed crops. Grazing on permanent grassland is also very common.

Farm mechanization is based on 25- to 75-horsepower tractors which are typically foreign-made, and implements of Dutch manufacture adapted to local crops and conditions.

The Dutch climate is mild and damp, and greatly influenced by the sea. Temperatures are never extreme; snowfall is rare and rainfall is evenly distributed during the year. The cropping season is about nine months, beginning in March.

In the **Dutch Polders** (polders refers to drained land claimed from the sea), farmers rent their land from the government on long-term agreements, which encourage the best possible farm management. In spite of Holland's very dense population, the Dutch use space cleverly and are a net exporter of many common crops. The average cattle production of 115 dairy cows per 220 acres is also quite remarkable.

Intensive animal production in climate-controlled housing for poultry, pigs, and calves is also well advanced and commonly used.

Many common horticultural crops such as vegetables and flower bulbs are produced by farmers, and these special crops account for about 20 percent of the total farm income.

France. France is the largest country in Western Europe and has a very diverse countryside, offering flat land, gentle rolling plains, and large mountains. Most of France is blessed with rich farmlands, making it Western Europe's leading agricultural country.

Agriculturally, France is also very diverse. The typical farm is in the 110- to 350-acre range, specializing in three to five crops and one to two different types of livestock, of which dairy or beef cattle and pigs are most common.

France's crops include wheat, barley, rye, potatoes, sugar beets, various oil seeds and forage crops for grazing, silage, and hay. In the Bordeax region of France, grapes are produced and are used almost exclusively for the production of wine.

French agriculture is well mechanized with 50- to 200-horsepower tractors, and combine harvestors and implements of European manufacture.

Other European Countries. Other countries in Europe are similar agriculturally to the countries already presented, with some notable exceptions. Germany, for example, is highly industrialized and produces only about two thirds of its food requirements. It imports most foodstuffs as well as large quantities of concentrate feed components from overseas to maximize domestic livestock production. Like Germany, Switzerland is only able to produce about 75 percent of its food needs; however, some specialty products are exported. Switzerland requires farm mechanization based on small tractors due to the many steep fields. Often the farm mechanization is specially adapted for Swiss conditions.

An Introduction to Selected Other Nations

Canada. Canada is the second largest country in the world and covers more than 50 percent of the North American continent. Most of the population lives less than 450 miles from the United States border, and most farm land is found there as well.

Only 17 percent of Canada's land is utilized for farming, and of that only about one third can be classified as good farmland. Most of the other two thirds is used for permanent grazing.

The climate is mainly continental with long, cold winters and relatively short but fairly warm summers. Winter grain crop varieties are not common, and the spring crops are often of the 90-day varieties suitable for the short frost-free growing season.

Crop yields in the prairie regions are low by world standards due to limited rainfall that also limits fertilizer application.

Field mechanization in these regions is based on U.S.- or Canadian-made tractors, many of which are in the 200- to 350-horsepower range with four-wheel drive and capable of pulling large implements. Most farms have self-propelled combine harvesters and trucks for grain transport. Many farms have well-equipped workshops capable of handling common farm repairs.

Australia. Australia is mainly a flat, dry, and thinly populated land. Only a few regions along the coast receive enough rain to support agriculture and large populations.

Only three percent of Australia's total surface is cultivated, but another 62 percent is used for rough grazing for the very large sheep and cattle stations, where up to 45 acres per cow is required.

Wheat is the dominate grain crop in Australia, and on nearly all livestock farms, hay is produced from grass or oats. Sheep are the dominating type of livestock.

Farm mechanization is generally based on 100- to 200-horsepower four-wheel drive tractors pulling large implements. Large combine harvesters and other self-propelled equipment used are mostly made in North America, or under license in Australia. Most farms also have their own trucks for grain hauling and other transport tasks.

Due to the normally stable weather pattern, both seeding and harvesting seasons are longer than elsewhere. Being a southern hemisphere country, Australia's seasons are opposite to Europe and North America. Nearly all grain crops are based on the winter varieties. This means that the bed preparation and seeding takes place mainly during April to June, with harvest during November to January.

South Korea. This nation has rapidly developed an industrial base and is regarded by many as an industrialized nation.

Korean agriculture is characterized by small-scale family farms averaging only one hectare, equal to the so-called backyard farming of some other agricultural exporting countries. In addition, more than half of Korean total farm income comes from rice crops, and off-farm income accounts for about 43 percent of the total.

Korean farmers are far more dependent on agricultural income for their livelihood than are their counterparts in Japan and Taiwan.

With these frailties, Korean farmers are experiencing enormous hardships in coping with agricultural import liberalization.

The Commonwealth of Independent States (The Former U.S.S.R. and Its Republics). The U.S.S.R. (now the Commonwealth of Independent States) was a vast country composed of 15 republics, each with its own native language and customs. Russia was one those republics—far and away the largest—extending from the central European section beyond the Urals to Siberia. Russia is now an independent state.

Although most of the former U.S.S.R. is located above the 45th parallel (north of Minneapolis), there are climatic zones that range from arctic permafrost in the North to subtropics in the Georgian republic to desert in central Asia.

Agriculture in the former U.S.S.R. defied classification. Major crops vary radically from region to region. Wheat and rye are major grain crops. Flax and cotton are common in central Asia. Fruits and vegetables are grown south of the Caucasus.

Almost all agriculture in the former U.S.S.R. was collectivized. State farms and collectives ranged in size from 1,000 hectares to 10,000 hectares, and usually included one or several villages. The director or manager was responsible not only for agricultural production, but also for education, health care, law enforcement, and cultural programs.

All of the Republics of the former U.S.S.R. are now independent states and most of them have entered into an agreement to form the "Commonwealth of Independent States." All are striving toward the possibility of private enterprise and leasing or selling land for agricultural purposes. There is an increased interest in, and demand for, commercial farming; however, many initiatives are needed to achieve this agrarian reform. These initiatives must be timed and implemented consistently with one another and with a macroeconomic policy that assures more balanced fiscal and financial management. Those responsible for agrarian reform will encounter many critical challenges.

These recent changes have evolved since 1985. At that time, Mikhail Gorbachev became head of the Communist Party. His program of "perestroika," or restructuring, aimed to bring new life to industry and agriculture by removing powers from the central government and letting people in the factories and on the farms make decisions. Gorbachev also introduced "glasnost," or openness, that welcomed cooperation with other nations.

U.S. Secretary of State James Baker, in a speech on European policy at the Aspen Institute in Berlin, Germany on June 18, 1991 stated, "The United States, for reasons of history, has a special role to play in supporting the process of change in the Soviet Union. As the Soviets demonstrate the will to help themselves, to follow President Gorbachev's call in Oslo to 'stay the course' on perestroika and the new thinking, then we can and should join them step-by-step."

Selected Developing Nations

Haiti. Haiti, the most densely populated country in the Western Hemisphere, is also the poorest, having per capita GNP of only about $370. Sixty-five percent of the 600,000 farm holdings average slightly less than 2.5 acres in size, while over half of the land area has a slope greater than 40 degrees.

The almost run-away rate of population growth has brought about a crisis in satisfying nutritional requirements. Increasing demands require more acreage at the expense of cash crops, and the denuded hillsides have resulted in excessive erosion and consequent massive loss of soil fertility. Over two and one-quarter million acres have lost all topsoil and become a barren wilderness. Declining yields of food crops, mainly corn, sorghum, and rice, along with depressed prices and government taxation and price-fixing, have combined to lower even further any incentive which farmers might have toward better farming and care for their land. Even the larger land holdings, characterized by absentee ownership, are experiencing lower profitability and regressive deterioration of lands.

Brazil. Almost everything about Brazil is big—big land, big forests, big rivers, big cities, and big debt. Brazil is the largest nation in South America, and covers nearly one-half of the continent.

Part of Brazil's plateaus and basins are covered with grass and shrubs that are excellent for cattle ranching; however, most of the basins are covered by rain forests so immense they could blanket much of the U.S.

Millions of plant and animal species live in the Amazon forest, many of them barely known to science. Unfortunately, however, the forest is in danger. Over the years, great swaths have been cut down, burned, and flooded so that farmers, ranchers, and miners could move into the interior. (In one recent year, a forest bigger than Belgium went up in smoke.)

Brazil's farms grow a third of the world's coffee and much of the world's cacao, soybeans, sugarcane, and oranges.

Nigeria. One out of every four black Americans is Nigerian. Known as the giant of West Africa, Nigeria has more people than any other country on the continent. With its high birthrate, Nigeria may soon become the world's fourth most populous nation after China, India, and the former U.S.S.R.

Many Nigerians dwell in mud-and-thatch huts in small villages where they farm, fish, or herd livestock.

Among the crops raised are sorghum, millet, peanuts, and cotton. Nigeria's moist southern regions support dense tropical rain forests—much of which farmers have cleared to plant cacao, rubber, and oil palm trees. Coconuts are also grown.

Kenya. The East African country of Kenya exhibits a wide variety of natural conditions ranging from desert to mangrove swamp. Two-thirds of the country is arid, and agricultural production features grazing lands supporting beef production; however, infestations of the tsetse fly severely limit raising livestock in certain areas. Much of the northern area is either mountainous and forested or within the country's national parks.

Food production crops are predominately corn (maize), wheat, millet, and sorghum. Major export crops grown include sisal, cashew nuts, coffee, and tea. There are four important crops classified under "**industrial crops**," consisting of agricultural commodities that are needed in the processing and preparation of other finished commodities. These include tobacco for the cigarette industry; sugar for the baking and confectionery industries; cotton for the textile industry; and barley for the brewing industry.

With a present population of 20 million, and a rate of increase of 4.3 percent, the relatively advantageous position of the country in terms of a well-fed citizenry who enjoy improved social conditions may not be able to be maintained. With the highest percentage rate of population increase among African countries (perhaps one of the highest in the entire world), birth control measures must be considered. An agricultural policy which places emphasis upon assuring small farm holders of continuing support by their government has been a positive factor in the stabilization and growth of the agricultural industry in Kenya.

Bangladesh. Fertile soil and ample water, the country's chief resources, enable Bangladeshi farmers to harvest as many as three rice crops a year. This is seldom enough, though, to feed the inhabitants of the world's most densely populated agricultural nation. Bangladeshi farmers earn money by planting jute, the country's chief export. Cash crops of tea are grown in the hilly regions of the east.

The rainy season is a crucial time for Bangladeshi farmers. When the rains come late, poor harvests cause widespread hunger. Bangladesh depends on foreign aid for survival. Its leaders strive against huge odds to improve the agricultural economy and to provide health, education, and employment for its ever-growing population.

The Great Ranges in Productivity Among Nations

The question is often asked, "Why don't farmers in other countries produce food as we do in the United States?" The most accurate answer is that in all developing nations (and some developed nations) there has been a certain degree of reluctance to direct needed resources into the agricultural production sector.

The agricultural production sector comprises over 60 percent of the population of the developing countries, which have the lowest per capita income. Policy makers have ignored this until recently because industry was regarded as the best bet in the development race, and it was thought that agriculture would ultimately benefit progress. The role of agriculture was to provide a surplus for investment in industry and to release labor to work in the new factories; farmers were not regarded as requiring incentives to produce.

Contrastingly, in the U.S., simultaneous rapid expansion of industries and manufacturing, the resources of education, ample research, and credit facilities were directed to the agricultural sector. This action, along with favorable climate and fertile soils, brought about highly productive agriculture.

The result of this unprecedented outpouring of advanced technology resources has become known as the American "agricultural miracle." The average yield of corn, our number one crop, after remaining between 22 to 26 bushels per acre for 140 years up to about 1930, had increased to 80 bushels and more per acre in recent years. The yield of potatoes had approximately quadrupled, and the yield of wheat and soybeans had doubled. There are similar records of increased production efficiency for broilers, turkeys, beef, and pork. Production of milk per cow and eggs per hen has almost doubled in this period.

Yield in the U.S. has continued to increase. By comparison, production in other countries is not as bountiful.

The average yield of corn in the United States is approximately 5.5 metric tons per **hectare** (1 hectare = 2.471 acres). By comparison, the average yield of corn (metric tons per hectare) for some other countries follows:

Italy	5.6	France	3.9
Canada	5.4	Yugoslavia	3.8
Hungary	3.9	Poland	3.5
Egypt	3.7	U.S.S.R. (the Commonwealth of Independent States)	3.1
Romania	3.4	Argentina	2.2
South Africa	2.3	Kenya	1.5
People's Republic of China	2.2	Indonesia	1.1
Brazil	1.6	India	1.0
Mexico	1.2	Philippines	0.8
Malawi	1.1	Nigeria	0.8

The great range in productivity indicates a corresponding potential for increased yields in much of the world, particularly in the developing countries.

It must be recognized that in addition to a favorable allocation of resources enjoyed by the agricultural sector, other factors have also played a part in making high agricultural productivity possible in the U.S. Favorable climate, topography, fertile soils, and a relatively well-fed populace are among them. That few other nations enjoy such an abundance of resources to provide food for their populace can be readily surmised in these statements by Spitzer (1981, p. 65):

> Today, less than a dozen nations produce more than ample food for their own people.
>
> Approximately 50 nations are essentially food self-sufficient. This means that the rest, or nearly 120 nations, are dependent on outside food sources.
>
> With most national populations growing geometrically, and man's wants, needs and desires all changing, our basic resources are under tremendous pressure. Land and its product, food, are top-priority commodities.
>
> The great adaptability of the human race to a wide range and variety of land types, rainfall, and temperature has characterized world agricultural development.

Theoretically, a specific proportion of the earth's land surface is available for the use of each inhabitant. Peoples of the Far East and China have by far the fewest hectares of land per person. Both North America and the Commonwealth of Independent States (formerly the U.S.S.R.) have more than adequate amounts of land. Africa and the Near East have total hectares per person almost equivalent to North America and the Commonwealth of Independent States; however, much is jungle or desert land not suitable for cultivation.

The expanding population requires that each hectare of agricultural land must support more people each year. In 1970 there were 82 people for each 100 hectares, today there are approximately 112 people for each 100 hectares. Behind these global averages, the concentration of people on the land base varies widely among the regions. For example, there are only 5 people per 100 hectares in Oceania while there are more than 500 in East Asia. When only the land used to grow crops is considered, the ratios are much higher—ranging from 50 people per 100 hectares in Oceania to nearly 1,000 in East Asia. Obviously, the pressure on the cropland resource is severe in some countries.

Figure 4–1 illustrates the uses of global land area. Of the global land area, 35 percent is utilized for agricultural purposes. Within this 35 percent, 24 percent consists of pasture and meadowland, 1 percent consists of permanent crops, and 10 percent is arable.

The Development and Maintenance of Farming Systems of the World

The term "farming systems" refers to an ordered combination of crops grown or livestock raised, the methods of husbandry used, and the practices commonly followed

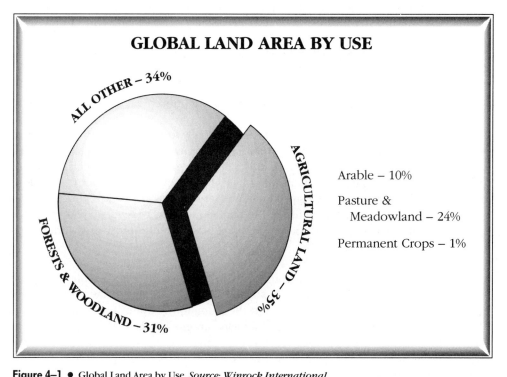

Figure 4–1 • Global Land Area by Use. *Source: Winrock International.*

by producers. If the question is asked, "Why is this specific crop grown or why is this species of livestock raised in a certain country?", the answer most certainly is that conditions and situations prevail in that particular locality which enhance and make profitable such commodity production.

Duckham and Masfield, in their extensive volume, *Farming Systems of the World,* presented a classification consisting of:

- tree crops.
- tillage crops (with or without livestock).
- alternating tillage (with grass, brush, or forest).
- grassland or grazing of land.

They further divided each of these four groups into subclassifications of (a) very extensive, (b) extensive, (c) semi-intensive, and (d) intensive.

The particular farming system found in any country, region, or district is the result of a unique combination of factors, many causative, others associative. Simply stated, there are fundamental reasons why bananas only thrive and produce food in specific locations while hard red winter wheat requires an entirely different, yet just as specific, location. In a similar manner, certain species of livestock are restricted to specific areas of the world if their husbandry is to result in economic production of food.

A listing of some of the basic factors and conditions responsible for the development of farming systems and the continued presence of crop and animal species in any particular location include:

1. effective climate (temperature, rainfall, and altitude).
2. land—physical form (terrain and soil types).
3. land—biotic or bio-geochemical status (fertility, organic matter, organisms, and pH).
4. moisture control facilities (irrigation, drainage, and floods).
5. soil stability (erosion and salinity).
6. unwanted species (weeds, predators, pests, and diseases).
7. technological adaptation feasibility (machinery, chemicals, and usage).
8. economic infrastructure (markets, prices, wages, government controls, and subsidies).
9. social infrastructure (dietary customs, work ethic, class status, rigid traditional practices, land tenure, educational levels, knowledge, and dissemination status).

While each of these determinants must be considered to some degree as basic to the types of farming in a given area, changes in one or more will often either expand the area where a given system may be found or, conversely, restrict the extent of the types of systems operating in the area. Let us examine a few of these areas more closely.

The Level of Technology*

Improvements in agricultural technology overcome the constraints imposed by few resources. If land is scarce, technology can provide seeds, fertilizer, and irrigation to increase the output. If labor is scarce new technologies replace labor and raise output. Developing nations have a potential to respond to appropriate new technology as they may use technologies that have existed for centuries. Research is the source of most new technology, and an extension service spreads new ideas to the farm population.

Weather*

Weather on food production is critical in all countries. Output is often determined by the timely arrival of precipitation as rain-fed agriculture prevails. Weather extremes such as excessive rains or drought impose hardships on agricultural production. The problems found in the Sahel region of Africa illustrate this.

Natural Resources*

Physical resources should be viewed as fixed, though their availability has importance to food output. Good soil, water, and a moderate climate that permits year-round crops are a few examples of the advantages of natural resources. All countries have natural resources and productivity can be increased by the implementation of development programs.

Infrastructure*

Food production is influenced by several aspects of the physical and institutional **infrastructure** available to farmers. Included are transportation, communications, electricity, roads, and storage facilities. Additionally, land distribution and leasing arrangements, availability of credit, seeds, fertilizer, water for irrigation, and the efficiency of the marketplace all affect the profitability of farming and the willingness of farmers to produce. Compared to the U.S. lesser developed countries have not developed these kinds of infrastructure and institutions. These are so fundamental as to represent the most critical needs. Without roads and economic forms of transportation, products cannot be moved profitably to the market for sale.

Production Incentives*

Production decisions are based mainly on the rewards producers receive for their efforts. The relationship of product prices and input costs, along with technology, production risk, and other factors, yields incentives or disincentives to producers. Farmers in the lesser developed countries are no different than farmers in the U.S.; they

---

*Adapted from the National Association of State Universities and Land Grant Colleges.

maximize the returns to their family even though more food is consumed in the home and less is sold in the marketplace.

Political Constraints*

Public policies that maintain low food prices for consumers are a primary influence on producer incentives, but do not motivate farmers to produce more. Public policies on exchange rates, import/export controls, and input prices are examples of other factors that influence farmer incentives.

Human Resources*

Agricultural production is influenced by the quantity and quality of human resources. The importance of labor in agricultural production has long been recognized, but more recently the quality of labor has received more attention.

People make land and other resources productive. Most developing nations have yet to provide enough training facilities and opportunities to create highly skilled workers. Illiteracy is high in many countries, skill training low, and public schools inadequate. Studies have suggested that the acceptance of agricultural technology is higher with farmers who have had more education. This does not mean, however, that an uneducated farmer is a poor farmer.

Nature and Extent of Farming Systems

Subsistence Farming Systems

Subsistence farming systems grow only enough food and fiber for their own needs, collect fuel and building materials from natural sources, and hardly enter into the cash economy at all.

Between 30 and 40 percent of the world's people and about two thirds of the world's farmers are living on intensively operated subsistence farms. Iraq, Egypt, and the countries of southern and eastern Asia contain a large proportion of the intensive subsistence agricultural systems.

All systems of subsistence agriculture are characterized by high inputs of manual labor relative to the use of labor-saving machinery and are thus characterized as labor-intensive operations, in contrast to the energy-intensive operations of almost all commercial farms.

---

*Adapted from the National Association of State Universities and Land Grant Colleges.

Commercial Farming Systems

The **commercial farming systems** of the world are characterized by the production of agricultural commodities for sale. In commercial agriculture, much use of capital is made for the purchase of tractors and machinery, fertilizers, insecticides, pesticides, herbicides, improved implants, better breeds of animals, and other technological innovations.

Types of commercial farming systems include:

1. tropical and sub-tropical plantations.
2. mid-latitude grain farming.
3. vegetable and fruit cultivation.
4. mixed crop and livestock farming.
5. livestock ranching.

The large amount of capital required for many kinds of commercial farming tends to greatly restrict ownership and favor the operator of large-scale units.

Collective Type of Farming Systems

Collective farming systems are usually located in nations that have a centrally planned economy. Centrally planned economies are utilized particularly by the countries of mainland China, North Korea, Vietnam, and Cuba. In such societies agricultural production often operates under a rigid system of collective and state farms. Some private ownership is allowed, and only rarely is group ownership permitted. In some countries, small private plots are provided for members of collective farms, state farm employees, and certain other workers. Reportedly, systems of state farmers and collective farms are not as productive as might be expected of comparable privately owned operations. However, one must recognize that under the collectivist system of mainland China, production and distribution of foodstuffs has increased and stabilized to the extent that the periodic famines no longer occur.

Extent and Distribution of Crops and Animals Providing the World's Food

The global land mass, when classified according to use, is roughly divided into thirds: 35 percent is used to produce agricultural products, 31 percent to produce wood and timber, and 34 percent in all other uses. The proportions are not constant because there is shifting from one category to another from year to year. Two-thirds of the agricultural land is used to support animal agriculture, 30 percent to produce annual crops, and only 2 percent to produce perennial crops.

Three large regions—sub-Saharan Africa, Latin America, and the Commonwealth of Independent States (formerly the U.S.S.R.)—contain almost one-half of the agricultural land, while European regions have less than four percent.

The total area in the world being used for field and horticultural crops is about 1.4 billion hectares (3.5 billion acres), out of a total of 3.2 billion hectares (7.8 billion acres) of potentially arable soils. Much of the unused potential, approximately 1.6 billion hectares, is to be found in areas presently lacking essential transport or other infrastructure needed for the maintenance of commercial food production.

Today, the world has abundant resources for producing food, fiber, timber, and industrial crops. Because of the affects of climatic change, migration, and social change, and in light of enormous changes in agriculture and its technology, well over 50 percent of the potentially arable soils of the world are not being used to any extent for farming. A large part of those being used could be used for larger harvests.

There are over 3000 plant species that have been used for food and more than 300 are widely grown; however, only a few species furnish 90 percent of the world's food.

Virtually all human nutrients come from three sources—crops, animal products (meat, milk, eggs), and aquatic foods. Of these three, crops are by far the most important. Crops furnish over 90 percent of the human population's calories and about two thirds of its protein. The remainder of the protein comes from livestock, poultry, and fish.

Among crops, the cereals (wheat, rice, maize, and millets) occupy three fourths of the cropland area and provide about the same proportion of the world's calories. Root crops, oilseeds, and sugar provide about 20 percent of the calories from crops. Vegetables and fruits satisfy only a very small proportion of calorie needs, but are very important for other nutritional values; see Figures 4–2 and 4–3 and Table 4–1.

Cereal Grains

A large proportion of the world's total food supply is provided by the group of plants classified as **cereal grains,** which include wheat, rice, corn (maize), sorghum, barley, oats, and rye. Also included are crops used specifically in one or a few countries such as "tef" in Ethiopia. Rice or wheat will generally be the leading crop, closely followed by corn (maize). Grain remains the cornerstone of modern food systems, the direct and indirect source of the bulk of the world's food energy.

Wheat is both grown and consumed largely in the temperate climate areas, while rice, more commonly grown in tropical and sub-tropical areas, is the staple food in countries making up half of the world's population. The people of China, India, and Bangladesh, along with those living in the countries of Indochina, depend upon rice to make up a large proportion of their diet.

The ten leading exporters of wheat in descending order are:

1. United States (28 percent)
2. Canada (22 percent)
3. France (16 percent)
4. Australia (15 percent)
5. Argentina (5 percent)
6. United Kingdom (5 percent)

Figure 4–2 • In Haiti, a man tends his family's vegetable garden using organic farming techniques taught by CARE. CARE also provided him with seeds. *Source: CARE; photo by Carlos Devillers.*

Figure 4–3 • Farmers in Kenya tend crops. CARE helps farmers worldwide grow more food by providing technical assistance, seeds, tools, and irrigation systems and by restoring land that has been damaged by soil erosion. *Source: CARE; photo by George Wirt.*

7. Italy (3 percent)
8. Germany (2 percent)
9. Hungary (1 percent)
10. Greece (0.3 percent)

The leading exporters of corn in descending order are:

1. United States (63 percent)
2. France (10 percent)
3. Argentina (8 percent)
4. China (7 percent)
5. Thailand (5 percent)
6. Yugoslavia (3 percent)
7. Belgium/Luxembourg (2 percent)
8. Romania (1 percent)

9. Italy (0.6 percent)
10. Canada (0.3 percent)

Non-Cereal Crops

Non-cereal crops contribute heavily to the 3.8 billion people attempting subsistence, and include the potato, sweet potato, yam, cassava, bean, soybean, peanut, coconut, and banana. Sugar cane and sugar beets also provide carbohydrates for a large proportion of the world's population.

The production and use of cassava appears to be slowly but steadily increasing in many of the developing nations. While grain production is the major energy source for human consumption and livestock feeding, cassava is an important food in many tropical countries. Cassava is a root crop sometimes known as tapioca. Major producers of this crop are Thailand, Indonesia, Nigeria, Zaire, and Brazil. A sizeable part of Thailand's production is exported to the European Community (EC) for use in livestock and poultry feeds.

Among tropical fruits, the citrus are important food crops that contribute much to healthy diets. This is particularly true with regard to Vitamin C. Commercial citrus producing areas include the following: northern Mexico, southern U. S., southern Africa, southern Asia, and most of the continent of Australia. Bananas, plantains, and papayas are important tropical and sub-tropical crops that are grown commercially in many countries (with major cultivation occurring in Latin America, Africa, and Asia). The economic and nutritional significance of these crops is well recognized, and they constitute a notable portion of world trade. Just as with the citrus crops, a considerable portion of the total volume of production comes from units of large acreage such as plantations or corporate farms. However, there are also many small and subsistence farmers in these areas who have ready access to native varieties of these crops. Also worthy of mention are two condiment crops, peppers and ginger. Countries producing ginger commercially include China, Taiwan, Philippines, Fiji, and Malaysia. Pepper crops are grown in many countries in the temperate and subtropical zones.

Dietary protein shortages among many of the world's people, particularly in the developing nations, are a major problem. Soybeans and other legumes often supply much of the protein in the diet. However, livestock, poultry, and fish constitute a major portion of the protein nutrient requirement. This is particularly true for livestock and poultry in the more developed countries; see Figure 4–4.

Livestock

Animals are important assets because they integrate well in sustainable farming systems. Livestock play a vital role in meeting nutritional, agricultural, and economic needs of people. Animals provide vital protein which is lacking in many diets. Examples of the advantages of animals are as follows:

1. ruminants eat and digest plants grown on land not suitable for cultivation.

Figure 4–4 • A woman in Belize feeds chickens she learned to raise in a CARE agricultural training program. Participants also learn business management skills and receive loans to get started, helping them increase income and escape poverty. *Source: CARE; photo by Steven Maines.*

2. animals produce manure which helps improve soil fertility and crop production.
3. animals can be used as a living savings account, food reserve, and draft power.

Global meat consumption is characterized by three distinct patterns. In the industrialized regions, consumption is high and little growth in consumption will occur with rising incomes. In the upper-income developing nations, consumption is generally already above the world average, but growth potential is great. In the lower-income developing nations, greater consumption is severely constrained by lack of income. It is in these latter two groups, with their large populations, where most of the consumption growth will occur and create greater demand for grains and oilseeds for livestock feed.

Considerable criticism has been made of the continued use of meat animals in diets. Such criticism is based upon the premise that livestock are wasteful users of grains which could be better used consumed directly by humans. "Such implications incorrectly depict the situation in the U.S. and present an even more distorted picture worldwide" (Hodgson, October 1976, pp. 625–630). Ruminant livestock in particular can economically convert forages and nonedible products for humans into nutritious high-protein

Table 4-1 •

Origin of Common Domestic Plants

North America:	sunflower, tepary bean
Central America:	cacao, cotton, maize, papaya, scarlet runner bean, sieva bean, tomato, cassava, sweet potato, common bean
Highland South America:	lima bean, potato, common bean, cotton
South America:	avocado, peanuts, pineapple, cassava, cotton, sweet potato
Mediterranean:	cabbage, grapes, oats, olives
Europe:	buckwheat, raspberries, rye, sugarbeet
Central Asia:	alfalfa, common millet, hemp, foxtail millet, grapes
China:	cabbage, onion, peach, soybean
Southeast Asia:	banana, citrus, mango, Oriental rice, tea, yam, thin sugarcane
India:	cucumber, eggplant, pigeon pea
Near East:	barley, chick-pea, date, fig, flax, lentil, onion, pea, pear
Africa:	African rice, coffee, cowpea, figa millet, sorghum, watermelon, yam
South Pacific:	breadfruit, coconut, taro, nobel sugarcane

food. Proteins from animal sources have a higher biological value than those from plant sources because they are rich in amino acids, lysine and methione, and they contain the essential vitamin B_{12}.

Ruminant animals include sheep, goats, cattle, water buffalo, alpaca, and camels. Cattle, sheep, and goats constitute the greatest source of human food derived from animal sources worldwide. Monogastric animals such as pigs and chickens require concentrated feeds, including large amounts of grain and grain products. It is probable that the strains of a rapidly increasing world population may eventually dictate a reduction in numbers of monogastric animals (particularly swine) in favor of the use of grains for direct human consumption.

A considerable portion of the total meat consumed is in the form of beef. The production of beef cattle is worldwide. While India and Pakistan have the highest density of cattle per square mile, these are of negligible importance from the standpoint of meat production because to the Hindu, eating beef is not acceptable and is considered taboo.

The nine largest world producers of swine in descending order are:

1. China
2. Commonwealth of Independent States (formerly the U.S.S.R.)

3. United States
4. Brazil
5. Germany
6. Mexico
7. Poland
8. Romania
9. Netherlands.

Fish and Aquaculture

Another important source of protein is fish. For many people of Asia and the Pacific, fish is an important food staple, and fishing is their livelihood. The world demand for fish is expected to double by the year 2000.

Since the beginning of the era in which fishing has become commercialized and increasingly dependent upon more sophisticated equipment and methods, concern has risen regarding limits to the potential of fish as a staple source of world food supply. The recent development of massive fishing fleets by some of the industrial nations such as Russia and Japan may, in the opinion of some, be bringing us frighteningly close to an absolute limit on the size of catch that can be made if seed stocks are to be maintained. It must also be recognized that fish farming has certain limitations compared to crop production and the raising of domestic livestock.

In the very early days of agriculture, farmers collected wild seed and planted it, a practice that roughly compares with the state of **aquaculture** today. Because of the limitation that most fish will not breed in captivity, young fish must be caught along the shore; see Figures 4–5 and 4–6.

Milk

Currently, global milk consumption is highly concentrated. Consumption in Europe is several times the world average and is also high in North America. The ten largest world producers of milk, in descending order are:

1. Commonwealth of Independent States (formerly the U.S.S.R.)
2. United States
3. France
4. Germany
5. India
6. Poland

74 Chapter 4

Figure 4–5 • Inland fisheries, Jamaica. This joint United States Agency for International Development (USAID) and Jamaican government project is designed to provide a cheap source of protein for rural populations and to create employment and income for farmers. Pictured is the government crew sexing and sorting tilapia fingerlings at a fish farmer's brood pond. Female fish in the tub will be sold to a fishmeal plant. *Source: FAO; photo by F. Mattioli.*

7. United Kingdom
8. Netherlands
9. Italy
10. Brazil

Projected Growth Pattern of World Agricultural Demand

Presented in Figure 4–8 is the projected growth pattern of world agricultural demand, in millions of metric tons, reported in percent of growth from 1980 to the year 2000.

Figure 4–6 • The program for Integrated Development of Artisanal Fisheries in West Africa is run by the FAO and is financed by the Danish International Development Agency and the government of Norway. The program disseminates information and organizes training courses, workshops, and seminars. Project activities may vary from improving boats and gear to providing drinking water and health care. The program covers all aspects of fisheries, including research and technology. The picture depicts "sun-drying" as a common preservation technique in many fishing villages. *Source: FAO; photo by L. Callerholm.*

World Food Security and Overcoming Food Deficits

Limiting Factors in Food Production as Viewed by Some

Agricultural production is confronted with some glaring limitations. Our productive acres may be near full utilization. We are also becoming aware that our other resources, including those providing energy, are indeed finite. There is no way that present practices of scientific agriculture can even be continued unless plentiful sources of energy are available. We have also found that water and soil, basic ingredients for

Figure 4–7 • Rural housewife of the Phu Wiang watershed in northeast Thailand sorting silkworm cocoons. Locally produced silk thread has become an important source of additional income for the peasant farmers of this area. *Source: FAO; photo by Peyton Johnson.*

agriculture, have definite limits in the amounts of waste materials and chemicals they can absorb.

When the total amount of food produced in the world is considered, only a modest rate of increase has occurred. For both the developed and developing countries, world food output has been rising approximately 3 percent per year over the past 20 years.

World Agricultural Demand Patterns in 2000
(in millions of metric tons)

Region	Meat	Milk	Cereals	Oilseed	Fiber
North Africa/Middle East	13.2	43.0	142	10	1.7
Subsaharan Africa	9.9	19.3	108	9	0.8
European Community	25.4	111.0	133	44	1.2
Other Western Europe	7.3	26.3	58	10	0.4
The Commonwealth of Independent States	24.3	118.3	306	20	3.6
Eastern Europe	14.8	56.7	139	17	1.2
South Asia	4.1	72.7	291	17	6.4
East Asia	18.7	15.1	224	23	3.5
China, Vietnam, Laos, Kampuchea, North Korea	42.4	14.3	457	37	7.3
Oceania	3.2	8.8	16	2	0.4
Latin America	29.5	68.0	161	18	1.9
North America	33.9	79.1	254	43	1.5
WORLD	**226.7**	**632.6**	**2289**	**250**	**29.9**
Percent of Growth from 1980 to 2000	64	36	46	62	37

Figure 4–8 • World Agricultural Demand Patterns in 2000. *Source: United States Department of Agriculture.*

The current food problems are primarily in the developing countries. With the exception of African countries, food production rates have risen slightly; however, this will not be enough to bring many of the developing countries into the ranks of self-sufficiency. The inescapable conclusion is that food should be provided for the peoples of all countries, and increasing food production through a combination of research, training, investment in delivery systems for nontraditional inputs, investment for land development including water, marketing, transportation, and storage systems is necessary.

The ultimate limits on food production from all sources are set by the availability and quality of natural resources—chiefly land and water—and by weather. Food production levels are also determined by the supply of energy, the level of technology, economic forces, governmental policies, and customs.

Conclusion*

The concept of the supply of food is an economic notion not really understood. The supply of food is the amount producers are willing and able to produce at a given price. Producers supply larger amounts of food at higher prices and lesser amounts at lower prices. Agricultural production is not only a biological process but a technical one as well that combines biological production with economic forces and management.

Food production is not only represented by fixed physical elements. Such a view often concludes that the potential for increasing food output is small. Since there are few land frontiers to conquer, the world's best farmland is already being used and available water supplies are already being used.

This view ignores the importance of economic human forces. Land and water are important to food production, but as they become scarce, economic forces create strong incentives to use them more effectively. History shows how new technology and management skills have increased the production from land and water resources as scarcity has increased. Finally, the amount of food supplied is determined by several interrelated factors: level of technology; weather; natural resources; infrastructure; producer incentives; political constraints; and human resources. The possibility of increasing the future output of food is excellent. This view-point is based on two conditions:

1. the efforts of the past 35 years have provided a base of experience on which to build.
2. past programs have established a large technological base for agriculture that serves as an incentive for the future.

Discussion Questions

1. Identify the countries and regions where major production of basic food crops and animals are found.
2. How would you classify farming systems (from the standpoint of production) as they occur around the world?
3. On a map of the world, locate regions where the major food crops are grown. Where are the major food producing animals raised?
4. In what area of the world are fish an important staple in the diet? What are the limitations of fish farming?
5. Discuss the issue of animal vs. grain crops for economic food production.
6. Answer how to substantially increase world food production, particularly as this is related to:

---

*Adapted from the National Association of State Universities and Land Grant Colleges.

(a) arable land and water
(b) climate and weather
 7 Explain some of the differences between farms in European countries and how European farms differ from farms in Canada, Australia, and the Commonwealth of Independent States (formerly the U.S.S.R.).
 8 If a specific proportion of the earth's land surface is available for the use of each inhabitant, what is the ratio of productive farm land to each inhabitant?
 9 What is meant by infrastructure? Why is infrastructure important to producers of agricultural commodities?
10 Why is more than 50 percent of the potentially arable soils of the world not being used for farming?

CHAPTER 5
Environmental and Social Influences on Global Agricultural Production

Source: Delkalb Plant Genetics.

Terms to Know
ecological refugees
deforestation
industrial roundwood
slash-and-burn
desertification
salinization
water logging
sedimentation
urbanization
energy production
transportation
sustainable agriculture
ecodisasters

Objectives
After reading this chapter, you should be able to:
- explain the various types of constraints which are often associated with deteriorating environments.
- explain the relationship between the quality and quantity of available water and the potential for food production.
- identify and briefly explain how specific farming practices tend to either assist in maintaining soil fertility or bring about a loss of soil fertility.

- locate the areas of the world where rapidly developing environmental constraints are becoming critical.
- discuss the development of events in the social and political arenas in some countries which tend to create constraints upon the production of food.
- define sustainable agriculture and describe the objectives of sustainable agricultural practices.

Optimal Environmental Conditions

Each species of plant and animal has developed through the process of evolution such that there is a range of environmental conditions which must be recognized as optimal for the specific plant or animal. For each species there are outer limits of the environmental milieu beyond which the organism cannot survive or cannot thrive and reproduce. As an example, certain crop plants will have difficulty surviving when they are grown in highly acidic or highly alkaline soils. Plants also vary greatly in terms of the amount of water needed or the amount of water which can be tolerated. Likewise, animal species vary greatly in specific food and water requirements as well as air quality tolerances.

Farmers and livestock producers have learned that they can produce the greatest quantity and quality of foodstuffs only when they select, provide, and maintain conditions which are as near as possible to the optimal environmental conditions consistent with the needs of the particular animal or crop. These boundaries are known as environmental constraints.

The Effects of a Fragile Environment

For many of the world's poorer farmers it is a fact of life that the harder they work, the poorer they get. Their land is either too steep, too dry, or the soil too poor, to support more than a mean level of existence for a few people. Because of increasing populations and intractable patterns of land use and tenure, fragile environments are being subjected to more intensive use than they can sustain. Agriculturally marginal to begin with, they are becoming useless; their farmers, **ecological refugees.** Cropland is becoming more scarce with relation to pressures from world population growth, and soil quality is decreasing at an alarming rate in many countries, primarily because of improper usage and management.

Only 11 percent of the world's soil has no limitations on its suitability for agriculture. Of the remaining soils 6 percent is affected by permafrost, 10 percent is too wet, 22 percent is too shallow, 23 percent suffers chemical problems, and 28 percent is too dry.

Although the world demand for cropland is greater than ever before, the amount of cropland abandoned each year may also be at a record level. Some reasons for cropland abandonment, usually the product of economic pressures interacting with ecological forces, include:

1. desertification.
2. severe erosion.
3. waterlogging, salinization of irrigated land, and the diversion of irrigation water to nonfarm uses.
4. increased use of land for nonagricultural purposes.
5. pollution of soil, air, and water.

The future of world agriculture depends on the wise use of water and land resources. The quantity, quality, and location of ground and surface waters will have an important impact on agricultural production. Declining water supplies have combined with rising energy costs and low crop prices to reduce the economic viability of irrigated crop production. Surface water runoff and the possibility of pollution from agricultural, industrial, or municipal sources will be of great concern in the future. High priority water research, extension, and teaching programs include improving the efficiency of water use; improving irrigation application efficiency; improving water quality for agricultural, industrial, municipal, and recreational uses; weather modification; and surface and sub-surface drainage.

Current and potential agricultural land problems throughout the world include competition for agricultural land from competing uses, deterioration in land quality, and changes in land ownership. Soil resource inventories are needed to update soil surveys and to determine prime farmlands, productivity, erodibility, salinity, and other soil qualities. Information is also needed to measure long-term impacts of continued soil losses from wind and water on agricultural productivity and yields. Analysis of soil-water relationships, including the available water holding capacity of soils, is vital to improving agricultural land quality. The relationships among reduced tillage technology, crop yields, and herbicide and pesticide application levels must be determined under a variety of conditions. Furthermore, additional information is also needed to assess the impacts of changing agricultural ownership patterns on the future of agricultural and rural communities.

Increasing Severity of Environmental Constraints

Deforestation must rank near the top among the phenomenal culprits of increasing constraint upon world food production. The two principal causes of deforestation today are land clearing for agriculture and wood gathering for fuel. The world's forests are being cleared at the rate of 25 hectares (62 acres) per minute or somewhat in excess of ten million hectares per year—mainly for agriculture and logging. As a result, fragile tropical soils are rapidly deteriorating into marginal productivity, with some even becoming wasteland. Strongly associated with such change is a tendency toward decimation of indigenous tribal peoples and destruction of thousands of unique plant and animal species.

A definition of forests may be of vegetation complexes dominated by trees, but can also include brushlands, open woodlands, and other wildlands in a largely natural state. An important part of the agricultural industry are manmade plantings and the forest management practices employed in natural complexes and areas more intensively cultivated by man.

Industrial roundwood is the term used to designate harvested forests used as (1) sawlogs for construction materials, (2) pulpwood, (3) chips and particles, and (4) wood residues. Major production of industrial roundwood by country or region is tabulated as follows:

1. U.S. and Canada — 36.5 percent
2. Europe — 18.4 percent
3. Commonwealth of Independent States (formerly U.S.S.R.) — 17.9 percent
4. China — 6.1 percent
5. Brazil — 4.1 percent
6. Japan — 1.9 percent
7. All other countries — 15.1 percent

Use of forests for fuelwood varies a great deal from country to country. In terms of percentage of total harvested forests for firewood, proportions used are as follows: Japan 0.1 percent, Europe 16 percent, U.S. and Canada 17 percent, Africa an alarming 92 percent, Asia 77 percent, and Latin America 69 percent. Such statistics speak boldly of deforestation and productive land loss.

Forests also protect water catchments, reduce flooding, lessen wind erosion, and assist in maintaining agricultural production. Forest products contribute to agricultural world trade and are particularly important to a number of countries including the U.S., Canada, Brazil, China, Nigeria, Indonesia, Cote Divoire, Costa Rica, and Cameroon.

As many as one million species—up to one-fifth of the planet's total—could perish along with the forests by the year 2000. It is tragic to note that many varieties of plants and animals could disappear without ever being discovered by humans.

The value of some of the endangered resources is known: the camu-camu fruit from the jungles of Peru has 10 times the vitamin C content of an orange; the periwinkle plant found in the Caribbean is an ingredient for treating Hodgkin's disease and lymphocytic leukemia; and drugs for treating malaria and glaucoma originate in the tropical Amazon.

The increasing demand for tillable land and fuelwood in Latin America has threatened not only rare plant varieties but also whole forests. Central America, for example, was once blanketed with thick forests; today, less than half of it is forested.

Based in Costa Rica, the U.S. Agency for International Development's project on "Fuelwood and Alternative Energy Sources" has identified fast-growing trees that can be planted specifically to provide a ready source of fuel. Trials conducted throughout the region have identified thirty promising species of trees that are now being adopted by farmers.

Haiti has been particularly hard hit by deforestation. Haitian farmers cut down trees as a matter of survival. The U.S. is encouraging some farmers not to tear down naturally forested areas but to plant and carefully harvest trees as a cash crop on their own land. One creative program is providing Haitian farmers with plants that bear avocados and other marketable fruit.

The U.S. Agency for International Development's largest forestry efforts is in Asia. Projects in Indonesia, Nepal, the Philippines, and Thailand are seeking to establish systems of upland management that reduce environmental damage and incorporate sustainable agricultural practices. One of the longest efforts, based on nearly two decades of research in the Amazon areas of Peru, has centered on finding alternatives to **slash-and-burn** agriculture.

Worldwide, wood remains a vital energy source. Trees, branches, and sticks are all harvested as a valuable fuel source. Even in the developed world, firewood for heat continues to augment other fuels.

Worldwide vast areas of trees have been cut clear. Attempts to farm these cleared areas usually fail after a few years because of loss of valuable topsoil through **erosion.** In drier areas, deserts expand and the shelter of the oasis disappears. Even more critical, overcutting in many countries means no fuel for cooking. Some estimates suggest that as much as 30 percent of the world depends on wood for fuel.

The problems of erosion, expanding desert areas, silt-clogged rivers, and lower crop yields are the results of poor forestry practices; see Figures 5–1, 5–2, and 5–3.

Desertification, like deforestation, poses a major threat to even marginal increases in the level of world food production. More than one-third of the earth's total land area is comprised of arid or semiarid areas, the home of approximately 60 million people eking out a precarious existence. Deserts are growing at an alarming rate, with 60,000 square kilometers of land lost yearly. Such losses can be largely attributed to drought, overcultivation, overgrazing, and improper irrigation practices.

An example of desertification occurred when a long-lasting drought struck the Sahel in Western Africa. International news photos depicted suffering people and starving cattle in that parched land. The drought that struck the Sahel, the group of eight West African states stretching across the southern fringe of the Sahara Desert, lasted from 1968 to 1974. It was not the first time that drought devastated this area, nor was it the last. In 1991 the specter of extreme famine again began to stalk relentlessly across 16 African countries. With drought as a major cause, crops failed and water holes became empty. Conditions continued to worsen throughout 1992. 23 million people presently face starvation in eastern Africa, 30 million are at risk in southern Africa, and 7 million are at risk in western Africa. This disastrous situation is becoming recognized as the worst famine of the 20th century. Civil war and internal strife have also contributed to the magnitude of the disaster. The Sahel region, containing 30 million people in an area two-thirds the size of the United States, is situated in a precarious ecological zone of fragile soils and erratic rainfall; see Figure 5–4.

The U.S. Agency for International Development's satellite imaging technology has enabled the Agency to implement a "Famine Early Warning System." Therefore, USAID

Figure 5–1 • At a CARE-supported nursery in Ethiopia, men tend tree seedlings. Once grown, the trees will help reduce soil erosion and develop new sources of wood. *Source: CARE; photo by Santha Faiia.*

has been able to monitor whether or not a particular country would experience crop shortfalls. Extremely arid conditions in northern Sudan in Kordofan, in Darfur in the west, and in the central and eastern regions, resulted in almost total crop failures, as well as a 50 percent reduction in output by mechanized farms.

Salinization, waterlogging, and **sedimentation** also pose significant challenges in the struggle to prevent a continuation of the declining levels of food production

Figure 5–2 • A CARE forester in Guatemala inspects tree seedlings. Like many nations in the developing world, Guatemala suffers from a destruction of forests that erodes soil and threatens food production. *Source: CARE; photo by Rudolph von Bernuth.*

Figure 5–3 ● A peasant woman in the Gurung District of Nepal planting fast-maturing fodder and fuelwood tree seedlings on a newly worked terrace in the Himalayan foothills. The Nepalese government and the United Nations Development Program assisted by the FAO is attempting to integrate watershed management and to bring the population together in community forestry development. The United Nations involvement reinforces the Forest Ministry's activities which include the education and participation of local communities in planting quickly maturing fodder and fuel trees, and the building of erosion control structures. *Source: FAO; photo by F. Botts.*

already apparent in a number of Third World countries. Either too much or too little irrigation water can cause salinity, and undue accumulation of salts on or near the soil surface.

Erosion by water and damage from flooding constitute another major causative factor in the decreasing potential of lands for food production. A layer of fertile topsoil that took centuries to build can wash away in a rather short time when the land is completely exposed to the elements. Likewise, much fertile soil is lost because of flooding, both through the carrying away of large amounts of soil particles and sediment, and through the covering of fertile soils with sands and debris of low fertility. Population pressures have sent many farmers to steep mountainsides where, particularly with the slash-and-burn system often followed, severe erosion will oftentimes permanently destroy crop and pasture lands in a very short time.

Figure 5–4 • Food shortages threaten African countries in the Sahel. Pictured is "dune fixation" near Dire in the region of Tomboucton, Mali. *Source: FAO; photo by J. Van Acker.*

Increasing use of land for nonagricultural purposes is a reflection of the increase in world population and economic activity. These changes shift land use away from the production of food and put more demand on its use for other needs. These nonagricultural needs can be grouped into: **urbanization, energy production,** and **transportation.**

These mounting demands are now claiming cropland in virtually every country. In the U.S. surveys by the USDA have indicated that more than 2.51 million hectares of prime cropland has been converted to urban and related uses. Including the Third World, statistical projections indicate that by the year 2000 expanding cities will cover 25 million hectares of cropland. Squatter settlements are increasingly becoming a part of the life of most Third World cities, and greatly diminish the amount of cropland available.

Pollution of soil, air, and water is recognized as a growing constraint upon world food production. Foremost in terms of public outcry, particularly in the developed nations, is concern over the escape and disposal of toxic industrial wastes as well as air pollution, which at times may bring about acid rain contamination resulting in waters being unable to support aquatic life. A statistic found in the May 1982 issue of Development Forum asserted:

> According to the International Scientific Research and Technology Corporation, by the year 2000—compared with 1970 data—the amount of major atmospheric pollutants as well as of polluted solid wastes will treble and the quantity of contaminated water will increase by 330,000 million tons, because of world industrialization. Accordingly, an increase in adverse effects on human health is expected.

A matter of increasing controversy is the possible hazard accompanying certain herbicide and pesticide usage. Some are especially critical of regulations which may ban usage of specified chemicals in the U.S. and other industrial countries but allow these same companies to sell questionable products in developing nations. Of particular concern worldwide is pollution that threatens the use of fish for human consumption.

The combined effects of sewage, industrial wastes, heavy pesticide and fertilizer runoff, and oil spills have been increasingly visible over the last few years. Coastal fisheries often have suffered accordingly.

The Task of Maintaining Optimal Environmental Conditions and of Lessening Constraints

Unfortunately, in a majority of nations, environmental constraints are increasing in their severity, not only reducing present food production, but also intensifying reduction of food production potential. Deteriorating environments can be associated with so-called technological advancements; however, careful selection and application of technological practices can contribute much toward lessening constraints and actually establishing more optimal environmental conditions for plant and animal production. While additional research and experimentation are needed, enough is known to make increased food production a future reality in almost all countries. Actions which need to be taken include:

1. communications for getting essential information to, and ensuring it is understood by, farmers and producers.
2. development of national policies and practices designed to provide incentives to both industry and agriculture in order to foster adoption. These practices would ensure optimal environmental conditions for food-producing plants and animals.

Sustainable Agricultural Production

A definition of sustainability is to keep an effort going continuously, and the ability to last out and keep from falling. Such a definition would suggest that agricultural systems would be sustainable if production could be maintained at current levels.

The goal of **sustainable agriculture,** according to York (1989), should be to maintain production at levels necessary to meet the increasing aspirations of an expanding world population without degrading the environment. It implies concern for:

1. the generation of income.
2. the promotion of appropriate policies.
3. the conservation of natural resources.

Sustainable agriculture should involve the successful management of resources for agriculture to satisfy changing human needs while maintaining or enhancing the quality of the environment and conserving natural resources, see Figure 5–5.

Social and Political Constraints Upon World Food Production

Overpopulation of humans, animals, and plants often constitutes a serious constraint upon food production. Strongly ingrained traditions and customs in many cultures may inhibit increased yields even to the extent of enforcing taboos on the use of specific plant and animal foodstuffs. Policy and practices regarding land ownership and tenure may well operate as a formidable constraint, severely limiting production of nutritious edibles. Widespread apathy and lack of concern for the education of rural segments of the population in a number of developing nations shackles initiative and discourages production efforts among peasants and low-income farmers.

Government regulations and policies operate as constraints upon food production. At times, governmental decrees tend to undermine incentive for producers. Pressure to force producers to shift from food crops to crops for export is sometimes detrimental to both producers and the bulk of consumers.

Acceptance of conditions of poverty, malnutrition, and limited living conditions as an anticipated way of life for farmers and food producers also constitutes a serious constraint.

Finally, poor nations often find themselves in a vicious cycle. Large families create poverty. Poverty results in malnutrition and disease. High death rates encourage peasant families to produce more children in the hope that some will live. For these parents, a large number of children is an asset. The resultant high birth rate, in turn, inhibits the very social changes which might lead to prosperity.

Figure 5–5 • Two women tend crops in northwest Somalia, a region plagued by drought in recent years. CARE helps villagers improve their land through soil conservation and irrigation projects. *Source: CARE; photo by Rudolph von Bernuth.*

Conclusion

As nations and individuals around the world are made more aware of the severe deterioration of natural resources, both government and nongovernment organizations are called upon to implement the tasks of:

1. providing information about the relationship between technological change and environmental sustenance.
2. soliciting support of research efforts pertaining to environmental quality and control.
3. bringing about legislation and multinational agreements which will protect and restore the natural environment worldwide.

Figure 5–6 • In many countries, the environment is rapidly deteriorating. To counteract these trends, CARE teaches local farmers like this one in Niger that planting trees among crops can yield greater harvests while restoring the natural ecological balance. *Source: CARE; photo by Tom Sheffel.*

The environmental movement was shaken into action in 1968 by a series of **ecodisasters:** pesticide poisonings, super-tanker oil spills, smog, industrial water pollution, drought and desertification in the Sahel, deforestation of tropical jungles, and the rapid growth of nuclear power. Since 1968, the number of organizations working on environmental issues has mushroomed. In the developing countries for example, the number of organizations working on environmental issues has grown from fewer than 500 in 1970 to more than 3,000 today; see Figure 5–6.

Discussion Questions

1. Discuss at least six environmental constraints that tend to limit needed levels of world food production.
2. Discuss at least four major social constraints which tend to limit needed levels of world food production.
3. Explain why deserts are growing so rapidly and why our forests are becoming depleted.

4 Explain why there is an increasing use of land for nonagricultural purposes.
5 What recommendations might be implemented in order to better protect our environment?
6 Explain what is meant by sustainable agriculture. What are some possible outcomes of adopting the concepts of sustainable agricultural practices?
7 What are some reasons that seemingly productive cropland is being abandoned each year?
8 Why does the quantity, quality, and location of ground and surface water have an important impact on agricultural production?
9 Why does the practice of utilizing chemicals such as herbicides, insecticides, and pesticides result in controversy?

CHAPTER 6
Energy and Agricultural Production

Source: United Nations Development Program; Photo by Ruth Gassey.

Objectives

After reading this chapter, you should be able to:

- discuss the sources and forms of energy.
- explain energy systems as they relate to humans.
- discuss the variation between developed and developing countries in the distribution of energy usage in food systems.
- discuss actions that might be taken to relieve food producers of pressures resulting from a continuing energy crisis.

Terms to Know

organic
inorganic
energy intensive
fossil fuels
biomass
labor intensive
energy subsidy
hydropower

Classification and Uses of Energy Sources

Mankind, in order to survive and to engage in the activities which make life meaningful and productive, must have a source of energy sufficient to accommodate physical and mental needs. For humans, the basic energy source is food. The sun generates energy, and green plants capture and convert this energy into consumable form. The livestock we raise for meat and milk are also energy converters, but they too are fueled by plant products.

Energy sources can be classified into **organic** and **inorganic.** Organic sources are the result of the direct conversion of solar energy, which is captured by growing plants through the process of photosynthesis and stored as fossil fuels. The recovery and use of this long-stored energy from petroleum, natural gas, and coal is viewed as essential in maintaining the highly technical societies of the present world. In the U.S., fossil fuels supply 96 percent of the energy used for transportation, residential, and industrial purposes.

The term inorganic refers to energy created and released by action of physical phenomena such as water, wind, tides, geothermal heat, nuclear reactions, and direct solar light rays. Energy sources can also be categorized as being depletable and non-depletable; this is important in the allocation of research and development funding.

Fossil fuels, especially petroleum products, have provided substitutes for animal power and human power to accelerate the technological revolution in all of the developed nations. The U.S. alone (with approximately 6 percent of the world's population) is responsible for about one third of the world's total energy consumption. The U.S. currently requires almost twice as much energy as it did back in the 1970s, and will require almost three times as much by the year 2000.

The limitations of energy sources and supplies should be of serious concern to everyone. Even without any further industrial and technological growth and expansion, the supply of fossil fuels will be exhausted. It is disconcerting to reflect that the more affluent societies utilize much more energy than do lower-income societies.

Developed countries such as the U.S. are particularly vulnerable to effects of a fuel crisis. Of the energy we consume, only 4 percent is from renewable sources such as hydroelectric and solar. The remainder, or 96 percent, comes from the nonrenewable sources of petroleum, coal, and nuclear power. Seventy-two percent of this comes from petroleum.

U.S. agriculture is productive because of energy applied to the industry. Currently, the total food and fiber system consumes about 20.5 percent of the nation's energy. The make-up of this percentage is as follows:

- about 4.8 percent in processing of agricultural products.
- about 1.7 percent in distribution and transportation.
- about 2.8 percent in out-of-the home preparation.
- about 4.3 percent in in-the-home preparation.

- about 4.0 percent in fiber production (cotton, wool, and wood).
- about 2.9 percent in the actual production of raw agricultural products for subsequent use as food.

Although only about 3 percent of U.S. energy is used in production agriculture, this represents a significant part of agricultural production costs. Often the price of energy governs the type and intensity of agriculture that is practical. For example, many U.S. farmers are converting back to dryland farming rather than attempting to irrigate because of rising energy costs.

Another consideration is the timeliness of energy availability. Agricultural productivity is highly dependent on timely agricultural operations. A few days' delay in planting a crop can significantly reduce the yield of that crop. Furthermore, irrigation must be done when the crops need water, not when energy is available. A few days' delay in harvesting can result in loss of yield or crop quality. For example, it is estimated that a week's delay in a particular wheat harvest due to fuel shortages could result in a crop loss of 10 percent. In animal agriculture, there are reports of poultry producers losing as many as 4,000 birds due to a 30-minute interruption of power on a hot summer day.

Because U.S. agriculture is so vulnerable to oil import interruptions, attention must be given to energy conservation and to emergency measures that can be taken to ensure a food supply for the U.S. as well as for other countries of the world.

Adoption of energy-saving technology in the developed nations is constrained to those technologies which are economically beneficial if they are voluntarily adopted. For example, farmers are not likely to adopt energy-saving technology in the form of crop-rotation, tillage, or fertilizer programs unless they are convinced that it is to their economic benefit to do so. The adoption of alternative energy sources also depends on economic feasibility. Agriculture has the potential for producing energy in the form of **biomass** for combustion in the form of ethanol, methanol, methane, and vegetable oils. Programs to keep agriculturalists fully informed of these developments will occupy extension and teaching personnel for the benefit of world agricultural production.

One of the fruitful areas of energy research concerns energy conservation. Many opportunities exist, ranging from tillage practices through final processing and distribution of agricultural products.

Improved timeliness and precision of water, fertilizer, and pesticide application could reduce energy requirements. Currently, typical agricultural production practices in the U.S. and other developed countries are geared to average conditions. Fertilizer, irrigation water, pesticides, and tillage are applied uniformly to entire fields as though needed application is equal everywhere. Production and energy efficiency could be improved if methods were developed to tailor production practices to meet actual needs as they vary throughout a field.

This approach to specifying production practices requires improved knowledge of the economic and physical response of output to applied inputs and the natural environment. It requires equipment to sense existing field conditions rapidly, to make decisions automatically about needed alterations, and to adjust applications to meet this need—all while the equipment is operating in the field.

The need for multi-disciplinary cooperation is apparent. Plant and soil scientists, pathologists, and entomologists can better define needed inputs to accomplish desired results. Agricultural engineers are continuing to experiment with sensors, automatic controls, and equipment to quickly respond to changing conditions and provide the needed inputs. Currently, the feasibility of the entire process is constrained by economics.

Energy Consumption by Food Systems in Developed Nations Compared To That in Developing Nations

Energy use by food systems in the developed nations is not only much higher than in the developing nations, but also varies greatly in percentage consumed directly or on the farm, in the processing industry, and in commerce and homes. In the U.S. the processing industry uses the greatest amount of energy, commerce and homes the second largest, while farm operations use the least. When labor is applied as a function of the energy supplied to the food system, a tremendous decline is seen from 1930 to the present. The continued application of technology has brought about a situation in which, from the standpoint of farm production, one farmer in the United States now feeds approximately 80 people (some have estimated as many as 120 people). This increased energy usage through application of technology and labor replacement is known as the **energy subsidy.** In less developed nations, far less labor has been replaced by technological applications. Consequently, total energy subsidies for food crops grown are much lower.

If the developing nations are to be successful in substantially increasing production, energy for the production, distribution, and application of fertilizers, herbicides, pesticides, and machinery must be forthcoming. This must be accomplished with ever higher prices for fossil fuels on the world market.

Sources of Power in Developing Nations

In contrast to the intensive-energy systems of the developed countries, developing countries use extensive hand labor and animal power for crop production. One half of the world's food supply is produced with no animal or machine support. The stick, hoe, and human energy produce a great part of the world's food needs. Availability of power in the developed countries far exceeds that of developing countries. In Europe, there are 0.93 horsepower available per hectare of arable land, and 1.02 horsepower in the U.S. In the developing countries, however, the existing available power from all sources (human, animal, and mechanical) is estimated to be only 0.05 horsepower per hectare in

Africa, 0.19 in Asia, and 0.27 in Latin America. Additional inputs of human labor and draft animal equipment in these countries are expected to contribute more to the pool of power.

Traditional fuel for power sources, which still consume a considerable portion of total fuels used in developing countries, must include charcoal from wood, which is becoming more scarce because of the ever-increasing destruction of forests. The widespread use of animal dung for fuel also removes an economic source of fertilization for already marginally productive soils.

While most of the world must adjust to soaring oil prices, developing countries without their own oil are hardest hit. They are now spending $50 billion a year to buy oil—almost twice the amount they receive collectively from all outside sources for development assistance. At the same time, the world's poorest, most of whom rely mainly on firewood and agricultural waste for fuel, face another energy crisis—dwindling supplies of firewood. This combination is aggravating already severe economic and ecological problems.

Other forms of energy used by some developing countries include:

1. wind power, collected by mechanical devices.
2. direct use of solar energy.
3. **hydropower,** energy from falling water.
4. biomass, energy from anaerobic digestion.

It should be recognized that each of the four forms listed above constitutes a natural, renewable source and represents a highly desirable form. However, at present, the costs of construction and equipment needed to make use of these forms tends to put them almost beyond reach of the developing nations. Assistance in obtaining necessary finances and in implementing such energy producing programs is often secured through loans from World Bank affiliates, through United Nations development programs, and through grants from Private Voluntary Organizations (PVOs). Because of the extremely high costs involved, nuclear power has become a remote dream for developing countries. However, a recent breakthrough in manufacturing small reactors may bring about change in the feasibility of nuclear power for Third World nations. On the other hand, the growing concern on the part of many citizens in industrialized nations regarding the safety of nuclear power production may inhibit consideration of such an energy source by most developing nations.

Low Energy Efficiency and Corresponding Output in Developing Nations

In addition to examining present usage of various forms of power and their potential for future use in the Third World, energy output and efficiency should be examined. For the millions of small farmers throughout the developing countries, chief sources of power for farming are animal and human power. Most of these work animals are poorly fed and lack strength for heavy tillage. Largely because of malnutrition and hand tools, the human labor force for small farm agriculture has a high input to output ratio.

The Western approach to this low energy efficiency and output is to prescribe replacing animal and human labor with mechanical power. From past experience in developing countries we know this is not the answer. While a case can be made for selective use of mechanical power, especially in areas where multiple cropping can fit into the farming pattern, research is needed to make animal power more productive through the development of improved implements and hand tools that make labor more efficient.

The emphasis in modernizing small farm agriculture should be on labor intensity. If the families of small farm agriculture are to integrate new agricultural technology into their farming practices, it will be essential that animal and human labor be more efficient and that much of the drudgery be removed from human labor. For many countries, changes in family living and work patterns are involved.

Energy-Intensive Compared to Labor-Intensive Economies

Societies which have implemented advanced technologies have been able to do so because they were able to secure vast quantities of energy fuels. Thereby, they have shifted human labor from arduous expenditures of energy into management of highly technological food production systems which use other energy forms. It must be pointed out, however, than when energy-intensive systems are introduced into the economies of societies which have little industry other than agriculture, masses of the populace are inevitably left unemployed or underemployed. Consequently, the aggregate national income is often skewed in favor of a small elite class who were able to somehow obtain sufficient capital to secure and operate energy-intensive technology, often at the expense of greatly reducing employment for hand laborers. Even within the agricultural sector, application of improved technologies of the Green Revolution, many of them energy-intensive, has often had an adverse effect upon the welfare of the poorest segment of a developing nation's society.

Nations such as Bangladesh, Niger, Burkina Faso, Jamaica, and Haiti that have no indigenous energy sources are the big losers when there is an energy crisis. Costs for their fuel and fertilizers have increased over ten times since 1973.

So far these countries have been able to stay afloat by borrowing heavily from the international banking system. But credit for many of those countries is about to run out because they are so over-extended financially. When banks will no longer loan money to buy fuel and fertilizer, what will they do? Who can they turn to? They have no indigenous energy. They have no technical or economic base to research alternate sources of energy.

Energy Use In United States Food Production

Some interesting facts are noted when looking at energy usage for food production in the U.S. (Table 6–1). With only one farm family producing the food for approximately 80 other people, the populace of the U.S. is almost totally dependent upon American farmers to feed them.

Over the years as technology has improved, the American farmer has used progressively more energy to increase his productivity. While only 3 percent of our total energy use is consumed on the farm in food production, our entire food system consumes 20.5 percent of the total energy consumed in this country. Of that 20.5 percent, three-quarters is supplied by the fuels which have become most scarce—gas and oil.

The field crop farmer is not the only large agricultural energy customer. Farmers specializing in raising poultry, cattle, and hogs also depend on the availability of energy. For example, modern hog confinement buildings must maintain optimum environments to best utilize feed energy and to protect the farmer's investments. Feed processing and handling call for sizable quantities of energy. Dairy farmers depend on electricity for year-round daily milking operations and for quick refrigeration of their product.

Each type of farming operation consumes significant amounts of energy for transporting its products to market, as well as for powering its trucks and other mechanized equipment.

Energy Reduction in Agriculture

Opinions vary concerning the feasibility of reducing the energy required for agriculture. While it is generally recognized that energy consumption could be reduced somewhat, the question is whether this is economically feasible. In the U.S., the cost would be in the form of inconvenience and somewhat higher prices for food consumers. Among a number of practices being advocated are as follows:

1. increased use of natural manures.
2. biological pest controls.

Table 6–1 •

Food Related Energy Consumption	
Households	30 percent
Wholesale	16 percent
Transportation	3 percent
On-Farm Production	18 percent
Food Processing	33 percent

3. more careful monitoring of pesticide and herbicide applications.
4. continued intensive investigation in the areas of plant and animal breeding, and testing and applications of research findings.

The close relationships existing among energy, prices, and hunger must be fully recognized and viewed with concern by all.

If energy prices rise, the rise in the price of food in societies with industrialized agriculture can be expected to be even larger.

Alternative Energy in Agriculture

Agriculture has traditionally provided food and fiber, but in the future may be called upon to produce fuel as well. Such fuel may be in the form of ethanol, methanol, vegetable oil, methane, and biomass for gasification or direct combustion. With improved technology, potential for biomass as an energy source is great. Assuming one ton per acre of biomass could be collected from wheatland, a potential energy contribution of 70 million gallons of diesel fuel equivalent would result. Solar and wind are additional forms of energy that offer potential for agriculture because of their dispersed spatial configuration. For agriculture to realize the benefits of these alternative energy forms, more efficient means of capturing, converting, applying, and transporting them must be developed. The implications are tremendous—in the developed as well as in the developing countries. Solar energy is currently most easily used as heat energy; however, it is also collected as electrical energy usable in many common applications. Solar energy is currently expensive but may be feasible to drive devices remote from power lines or to provide peaking power for irrigation. Wind energy is another example of how electrical energy may be collected. Again, wind energy can be used in remote locations and to relieve electrical demands in some situations.

Conclusion

All segments of agriculture interact with each other and other segments of the economy. It is not possible to assess the impact of converting an agricultural product to energy without considering the effect this will have on the output of other products. A change in energy use on an individual farm may affect not only the practice under direct consideration, but other aspects of that farming operation as well. No problem is more critical today to humanity than that of energy.

Discussion Questions

1   Explain the meaning of the terms labor intensive, energy intensive, and energy subsidy.

2 Categorize energy sources and list several sources within each category.
3 Compare availability of power as applied in terms of horsepower per hectare in Europe, in the U.S., and in the developing countries.
4 Explain why developing countries are so hard hit when there is an energy crisis.
5 What effects on society can be anticipated when energy-intensive technology is introduced into the economy of a developing country?
6 Is it, in your opinion, feasible or possible to reduce energy requirements within countries that have energy intensive technologies? Provide a rationale for your response.
7 What are some alternative energy sources? How can they best be utilized, in your opinion?

CHAPTER 7

Appropriate Agricultural Technologies

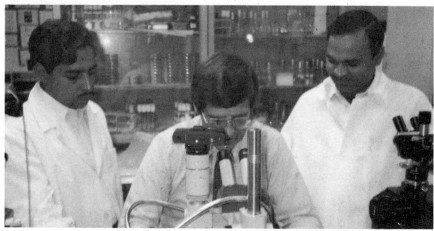

Source: Winrock International; Credit to Texas A & M University.

Objectives

After reading this chapter, you should be able to:

- explain the meaning of the terms associated with the various classifications of technology.
- discuss the ramifications of replacing labor-intensive technology with energy-intensive technology.
- explain why agricultural and construction manufacturing technologies developed for industrial countries may be inappropriate for other countries.
- identify some successful examples of the development of inexpensive agricultural technologies which are applied in the U.S. and abroad.
- list technologies which are of most interest to the developing countries.

Terms to Know

appropriate technology
technology-transfer
ecodevelopment
adequate technology
applied technology
intensive gardening
small animal husbandry

Technology Defined

One definition of technology—"a technical method of achieving a practical purpose"—is expanded into a sharper, yet broader, meaning by the addition of "the totality of the means employed to provide objects necessary for human sustenance and comfort." A more limited definition of technology is the "results from improvements in technical processes that increase productivity of machines and eliminate manual operations or operations done by older machines." Within this last definition is to be found our most common concept and usage of the term technology. It would be well to expand on selected meanings and aspects of the term. Technology is the relationship to the following:

- the way or manner in which a task may be accomplished.
- the devices or machines which are employed in accomplishing the task.
- the method used or pattern of activities followed in order to most effectively accomplish the task.
- a planned series of steps or activities which have been proven effective in accomplishing the task.
- the combination of resources and the extent and pattern of their use as inputs employed to bring about changes in production.

Christiansen (1992) defined **appropriate technology** as those sustainable, compatible, and affordable technologies introduced into or developed within a society that are deliberately adapted to fulfilling in that society's economic, political, physical, and social environment the needs of its people, with a minimum of unwanted social and cultural consequences. Additionally, Christiansen (1992) described technological change to be those alterations in technical procedures and products used by society to provide objects necessary in maintaining themselves, and which consequently affect the culture of that society, see Figure 7–1.

Technologies are often classified by their nature as being either energy intensive, labor intensive, or capital intensive. Other classifications express progression, hence categories such as primitive technologies, simple technologies, intermediate technologies, adequate technologies, modern technologies, complex technologies, and industrial technologies. Some of these terms may be used interchangeably. Categories suggested are not considered mutually exclusive.

Food Production and Technology

Workers in agriculture must be alert to technological viability and potential for application. Inadequate technology is a major reason why potentially arable land is not cultivated and why currently cultivated land produces such low yields. Several decades ago, many assumed that technology could be transferred wholesale from temperate

Figure 7–1 ● A man operates a traditional *shadoof* in a wadi near the town of Cheddra in central Chad. CARE has initiated a project to provide drought-stricken nomads and semi-nomadic people who have lost their herds and other food sources with an opportunity to switch to farming. Under the program, CARE has transformed several wadis, or dry seasonal riverbeds, into productive agricultural zones. CARE assists in selecting plots of land, providing seed and constructing water wells and 'Y'-shaped shadoofs which the farmers use to draw water from wells. A shadoof is a long pole on a fulcrum with a water bucket tied to a rope at one end and a counterweight at the other. Some 3,000 nomads and semi-nomads have been settled under this program. *Source: CARE; photo by Shaw McCutcheon.*

regions to the tropics. Much agricultural technology is transferable, but it must be modified to meet local conditions; this is particularly true for the biological components. New crop and animal production systems that are higher yielding and more profitable must be devised for thousands of combinations of soil, climate, and plant-pest complexes, as well as consumer preferences. New production systems must be devised for every crop for every season of every region and must be adopted by hundreds of millions of farmers, many of whom are uneducated and live in remote areas. The myth of general transferability of agricultural technology has been set aside, see Figure 7–2.

A few agricultural-related technologies which should hold considerable interest for many countries include the following:

- power sources and methods used in soil preparation, planting, tillage, protection, and harvesting of crops.

Figure 7–2 • These irrigation canals cross through tomato fields in Chad. The canals were dug by workers paid with food from the U.S. Food for Peace Program and administered by CARE. The project is part of a larger program to reconstruct existing agricultural and public health infrastructures within the country. Funds are being used to purchase construction materials and hand tools. The equipment is being utilized by laborers who repair small dams, water systems, irrigation canals, pumps, farm machinery, flood and erosion control embankments, fence construction, land clearing, and leveling. *Source: CARE; photo by George Worsky.*

- power sources and methods employed in the use of water, including irrigation, flood control, and reclamation of salinized soils.
- development, distribution, and usage of improved varieties of crops as well as breeds and strains of livestock.
- communication methods, technology-transfer techniques, and instructional aides employed by change-agents. The human resource has been and will continue to be the most important element in the development of agriculture.

Technology, Economic Growth, and Stability

The axiom that economic growth must be an overriding factor to consider in development is accepted with little challenge. Increase in gross national product (GNP) is accepted as almost mandatory for development and maintenance of an ascendant nation.

In purely economic terms, "development" has come to mean the capacity of a national economy whose initial condition is relatively static to sustain an annual increase in its GNP of between 3 and 7 percent while altering the structure of its product and employment so that a declining proportion of both is generated by agriculture, whereas an increasing proportion is provided by secondary and tertiary sectors. Monetary flows are likewise important; the outflow of hard currency for imports and expenditures abroad must not be substantially higher than the inflow of hard currency into the country from investments, loans, grants, and remittances.

Much of the increased efficiency of industrial countries has been brought about by the incorporation of labor-saving and time-saving technologies, particularly those supportive of mass production. The very size and scope of production in the Western world demands a capital-intensive technology base. Thus when the matter of **technology-transfer** from the developed nations to developing countries is considered, many problems soon become apparent. The constraints which are unique to developing nations have led many social scientists, both in developed and developing societies, to the conclusion that other factors besides economic impact must also be given consideration.

Organizations in developing countries have often been faulted for developing or promoting inappropriate technologies that are more suited to the economies of scale and technical sophistication of the developed countries.

It is a common observation that many of the developing nations, in their drive for modernization and development, utilize technologies that have been borrowed or slightly adapted from the developed nations. Furthermore, the result of having placed reliance upon the developed nations as a technological source has often been the installation of expensive, ill-adapted, and socioeconomically damaging technologies (damaging in the respect that surplus labor and indigenous natural resources are not optimally utilized). However, it should be recognized that some countries have done so under pressure from external, especially bilateral, technical assistance agencies.

These and similar constraints have resulted in much attention to the matter of designing technology that will account for the unique constraints affecting technological development. The more commonly referred term is appropriate technology, although such terms as intermediate, adequate, and modern have also been used.

Technology and Organization for Rural Development

Appropriate technology as applied to many elements of technology-transfer from industrialized western nations to countries of the Third World poses a dilemma. Such matters as impact on the society of the recipient nation must be given consideration. In particular, poverty increase resulting from unemployment must be given priority when considering plans for technological adoption and adaptation.

Economic development theories that held sway during the past are beginning to lose their credibility in light of mounting unemployment throughout the world. It is evident that a massive transfer of modern technology from rich to poor countries does not provide the key to prosperity in much of the developing world. Capital-intensive, energy-consuming, labor-saving technologies utilize the very resources that are scarce and expensive in many countries, while failing to utilize much of the country's most abundant asset—its people.

When a poor country invests a large proportion of national savings in imported capital-intensive technology, such action may often raise the productivity of a few workers, and consequently the GNP may be similarly increased. But, too often such action seriously depletes the capital available for aiding small farmers, landless laborers, and cottage-type small-scale industrial producer-workers, the very people who constitute the majority of the work force in most developing countries.

Besides the necessity of keeping the populace employed, human satisfaction and the attainment and retention of cultural values must be accepted as preeminent.

Let us examine the dangers of adapting technology without regard for its cultural ramifications. The elite in most newly independent countries have made costly mistakes for their peoples by adopting some Western models of development, education, and technology. The conditions in some of the newly independent countries were not conducive to unquestioned imitation of values, technologies, and institutional forms from Western societies. Rulers who took over the reins of government from the colonial powers were not an integral part of the culture and traditions of the newly independent countries. It was not surprising that the same sophisticated and capital-intensive technologies were introduced to those countries in ignorance of the realities of past civilizations, and of the present human resources.

Technology is only a means to an end. No sensible person will contest the need to provide the basic and essential needs of the population and of improving their standard of life by using technology. But people should think of the human cost of artificially creating endless material wants and of the craze for directing all resources and know-how to increasingly satisfying these wants. A purely materialistic approach to life will not take us to increased happiness or to the joy of living. Human life is capable of higher achievements.

The accusation is sometimes made that even the transfer of communication techniques, educational methods, teaching practices, and curriculum organization, as well as implementation of programs for nonformal teaching are possible examples of ineffective and inappropriate technology-transfer.

While the matter of technology-transfer of educational methods and communication skills will be addressed in more detail in upcoming chapters, the appropriateness of technology-transfer in the educational and skill training areas is of considerable importance in furthering development in any country.

Criteria for Evaluating Technologies Considered for Transfer

Over a number of years, certain criteria for use in predicting success for the transfer of a given technology have been proposed and established.

Ecodevelopment, for example, is an approach to development which integrates economic, social, and ecological dimensions.

Technology for ecodevelopment must be suitably adapted to any given situation—compatible with the characteristics of the particular population, the predominant ecosystems, and the local economic situation. It could be called **adequate technology** and would include the following:

- compatible with the productive subsistence of local ecosystems.
- simple to operate.
- low-cost in terms of investment and maintenance.
- labor-intensive and able to generate new employment when necessary.
- compatible with basic sociocultural patterns that would allow for popular participation, increase productivity and income, stimulate distribution of income and power, tend towards independence from foreign technologies, and increase self-reliance.

Basic to the appropriateness of any technology is careful consideration of how well such technology will contribute to or enhance the following:

- income level of each family.
- satisfaction of basic needs.
- resource conservation and development.
- societal development.
- cultural development.

- human development.
- environmental developments.

Examples of adequate or appropriate technologies which may be adopted by developing nations include the following:

- pedal power devices for application to irrigation pumps and threshing machines.
- wind power devices for application to various farm and household tasks. An example is a windmill so sensitive that it spins in breezes of under four miles an hour to pump nearly 4,000 gallons of water a day.
- water (hydrologic) power devices for generating direct or indirect (such as electric) power. For example, small hydroelectric generators that can obtain enough power from small streams to supply a farm or a school.
- solar power devices for water distillation and dehydration of foodstuffs. Another example includes a solar hot-water heater made out of burned-out fluorescent light tubes.
- small units for generation of biogas and alcohol from waste products and plant growth.
- improved tillage and harvesting tools made from local materials with local labor.
- new or improved varieties of plants and breeds of animals that are more efficient in food production.
- practices leading to improved nutrition for both humans and animals.
- effective communication and adult education programs.
- improved practices in production management.

All these innovations are designed to work without using petroleum for fuel.

Another example of appropriate technology as reported in the WORLD DEVELOPMENT FORUM (1984) was that in India, the bullocks and buffaloes haul more freight than the railways. It was also noted that India loses about a million animal-years of work each year because of necks ruined by the traditional yoke—apparently a case of bad design. The straight wooden beam touches only a small area of the animal's neck, which therefore bears the entire weight of the load. The wood digs into the flesh and itself adds an extra 85 pounds. Simply replacing the yoke with a modern horse collar increases a water buffalo's power by half. It was estimated that 400 million horses, oxen, cows, water buffaloes, donkeys, camels, mules, yaks, llamas, reindeer, and elephants contribute as much as 90 percent of the agricultural power used in some developing countries.

The example just presented, was noteworthy; however, let us examine a case where a team of experts makes recommendations with little local involvement. When a feasibility study for a cement plant in Africa was carried out in the dry season by a team of experts with little local involvement, the plant began operations in the rainy season and the company realized there was no equipment to dry wet limestone.

It happened in Benin, and similar foul-ups occur all over Africa. Relying on foreign research and technology can result in costly mistakes. Governments have to choose which technology to use in their projects. But with little understanding of technological issues, and faced by articulate foreign experts who press the best answer, governments often make choices that result in development failures.

Sometimes people with technical expertise offer advice without really getting a good feel for the problems a particular region is experiencing and may present for consideration a poor recommendation. The impact of this mistake is obvious. Short term consultancy has been and will be an effective method of providing assistance; however, consultants should take advantage of getting to know the people, the culture, the problems, and more importantly, learn to have vision pertaining to what implications might result based on advice given. (Refer to Appendix B for tips on becoming an effective consultant.)

Obtaining Practical Technical Expertise

The comments in the preceding section assume that a person has the technical expertise necessary to provide appropriate technological assistance. What if a person is called upon to provide technological assistance in unfamiliar areas? How is such expertise gained?

Students who have graduated from college are well oriented in theory and have received a very good general education. However, many of these students have not received instruction in **applied technology** and are inexperienced in the practical application of agricultural practices. More often than not, these students come from urban environments and have little, if any, exposure to the real world of agriculture. They are not alone—many agriculturalists have not developed skills in a particular area. For example, a high school agricultural instructor or a county cooperative extension agent may know very little about sheep. The question becomes "Where can I gain hands-on practical experience?" There are many responses to that question. They can attend workshops and clinics, or spend time working on a farm.

The sections that follow describe three organizations that provide opportunities to gain some specialized practical technical expertise in a variety of ways.

The HEART Program

The HEART (Hunger Elimination Action and Resources Training) program is a prime example for the teaching of topics relating to appropriate technology as applied to developing countries; see Figure 7–3. The HEART program is headquartered at Warner Southern College (Lake Wales, Florida) and has been developed to provide an opportunity for people to develop technical expertise in rural and urban development. Warner Southern is a four-year fully accredited Christian college of the Liberal Arts tradition.

The primary purpose of the HEART program is to educate persons through living and

Figure 7–3 • Agricultural skills training project in Jamaica. This government supported program through HEART is designed to train young people between the ages of 13 and 30 in agricultural skills with particular emphasis on the expansion of export crop production. Skills imparted are mainly in the areas of coffee, citrus, banana, coconut, and tobacco production, as well as horticulture for fruits and export vegetables. Training is given in practical and basic theoretical aspects of agriculture, including new techniques in farm planning, budgeting, post-harvest care, and other requirements for good agriculture. *Source: FAO; photo by F. Mattioli.*

learning experience including practical and technical skills and cultural communication. More specifically, the HEART program educates persons in the following areas:

- the skills of living as members of the community, meeting physical, spiritual, and emotional needs.
- the teaching of "grassroots" development as a model to improve the quality of life of the poor.

The curriculum of instruction is designed to address the most basic needs of the poorest of the world in a holistic manner. The curriculum provides a balance between theory and practice as the theories learned are put to immediate field practice. Components of the program include the following:

Intensive Gardening. Oriented to the Third World, this calls for appropriate methods and techniques of soil preparation, planting, irrigation and moisture control, appropriate technology, pest control, soil characteristics and conservation, and testing on small plots. Emphasis is given to under-exploited and tropical plants.

Small Animal Husbandry. This provides the student with both the theory and practice of raising small animals in a Third World environment. Special emphasis is given to feeding, breeding, and managing methods and techniques for raising small animals for food and resale.

Nutrition and Food Preparation. Stresses the principles of nutrition and preparation of nutritious and economical meals in a variety of cultural styles. Village-produced food is incorporated into menu planning and meal preparation.

Primary Health Care. Provides the student with essential disease prevention and health promotion information, first aid skills, and family health.

Appropriate Technology. Provides an introduction to technologies which conserve natural resources. It covers uses of hand tools, drilling of wells, sanitation technology, water systems, simple surveying, building materials, small machinery, water pumps, and irrigation systems. Also, construction and maintenance of brick, adobe, thatch, and bamboo structures are examined.

Cross-cultural Communication. Includes demonstrations of cross-cultural communication techniques for community development.

It is important to point out that HEART's training site is a simulated Third World village on a wilderness tract of land in sub-tropical Florida. Lacking most modern conveniences such as electricity and modern plumbing, the village is consciously structured to approximate Third World living conditions. The community is designed to be as self-contained as possible.

The following is an outline of the topics covered for the various educational topics just discussed:

Appropriate Technology. Topics covered include the following:

1. plain table surveying and tube level techniques. Used for moving water from one place to another and for construction.
2. health and sanitation for latrines and other sanitary facilities.
3. developing water sources and water system design.
4. water purification.
5. concrete construction.

6. cooking devices such as mud stoves and lorena stoves.
7. construction materials (adobe, wood, and bamboo) foundations, walls, floors, windows, doors, and roofs.
8. solar, wind, and water power applications.
9. use and care of basic hand tools.
10. small engines and how to keep them running.

Primary Health. Topics covered include the following:

1. helping health workers learn.
2. dehydration and fevers.
3. sicknesses that are often encountered.
4. how to examine a sick person.
5. nutritional issues in developing nations.
6. preventive health care.
7. medicine and medicine kits.
8. instructions for injections.
9. first aid and CPR.
10. illnesses that need special attention.
11. problems of the skin, eyes, teeth, gums, mouth, urinary tract, and genitals.
12. childbirth, mothers, midwives, and babies.
13. dental health and hygiene.
14. personal and family health.

Food Preparation and Nutrition. Topics covered include the following:

1. hindrances to the world food economy.
2. general human nutrition.
3. infant and child nutrition.
4. nutritional evaluation of diets.
5. recognizing signs of nutritional deficiency.
6. menu planning.
7. conserving nutrients during cooking and storage.
8. energy conservation during food preparation.
9. cooking basic staples such as beans, rice, cornmeal, powdered milk, and wheat flour.
10. rendering fat, canning, soap making, stir frying, baking, braising, pressure cooking, sprouting seeds.
11. preparing food from other cultures.

Intensive Gardening. Topics covered include the following:

1. philosophies of agricultural production.
2. land preparation.
3. soil fertility and management.
4. planting and transplanting.
5. soils and fertilizers.
6. insects and plant diseases.
7. weed prevention and control.
8. irrigation.
9. marketing and management practices.
10. cultural and management practices for vegetable crops.

Small Animal Husbandry. Topics covered include the following:

1. philosophies of agricultural production.
2. animal production and management for rabbits, poultry, goats, and fish.
3. animal nutrition.
4. pastures and other forages.
5. animal production in the tropics.
6. tropical feedstuffs.
7. feed ration formulation.
8. slaughtering of small animals.

Heifer Project International and the International Learning and Livestock Center

Heifer Project International (HPI) has its world headquarters in Little Rock, Arkansas, and the International Learning and Livestock Center (ILLC) is located at Perryville, Arkansas. HPI is a non-profit organization that supplies food- and income-producing animals, training in animal husbandry, and sustainable agriculture to low-income or resource-poor rural families in need throughout the world. The initial gift is repaid by an offspring being passed on to another family. The ILLC is a 1200-acre facility that uniquely blends livestock enterprises and learning experiences.

"Heifer Project International," a U.S. State Department official has stated, "is solving the problem of hunger around the world one farmer at a time." For over 40 years, HPI has provided livestock and training to low-income farmers overseas and in the U.S.

More than 90,000 improved farm animals and 1.5 million chicks have been distributed to needy families in 110 countries and 31 U.S. states to produce more food and added income. In addition, HPI has distributed more than 5,000 beehives and 1.5 million fish fingerlings.

Dan West, an Indiana churchman, founded Heifer Project with a vision. While volunteering as a relief worker during the Spanish civil war in the 1930s, West saw the rations of powdered milk quickly depleted and the impoverished families returning for yet another handout. Why not give a heifer instead, he thought, that will provide a lasting source of milk for a family? And this same gift can be passed on by the recipient, in the form of an offspring, to another needy neighbor.

Additionally, HPI offers a number of other resources to help others help themselves. Available from HPI are general technical brochures, current project news, suggested activities for schools, churches, and civic clubs, global awareness study materials, audio visuals, and books. There is a small fee for some materials, while other materials are free.

Purchased in 1971, "the ranch" was to serve as the holding center for the many animals that are donated and shipped to developing countries each year. Animals at the ILLC include:

- cattle
- sheep
- swine
- rabbits
- fish (fingerlings)
- draft animals
- goats
- poultry
- bees
- exotic animals

The Brangus cattle, Katahdin sheep, and Jersey dairy herd are production enterprises. These animals are managed in such a way as to maximize income so as to underwrite ILLC expenses. Over the years HPI has realized training potential of the ILLC for livestock management demonstration, giving a visible presence to promote HPI, leadership development, as well as putting on state, regional, national, and international conferences and seminars, see Figures 7–4, 7–5, and 7–6.

HPI and ILLC facilities include a complex of buildings to accomodate overnight lodging for groups as well as conference space for up to 100 persons. Housed there as well is a cafeteria, resource center, laboratory, and office space. Along the Educational Trail, visitors find working examples of the animal and agricultural techniques HPI helps communities around the world to develop.

The work of Heifer Project International is to a large extent made possible by volunteers. A few work directly in the project areas, others work to raise funds in their communities, while still others come and donate their time and talents. The relationship between volunteers and the ILLC is a mutually beneficial one. The volunteer gains firsthand experience of the work of HPI and the ministry of livestock development while the efforts of HPI are enhanced by their labors.

Figure 7–4 • Heifer Project International's country-wide dairy project in Tanzania, Africa, is the largest and one of the most successful of all HPI projects. In several African countries, dairy heifers are helping war widows and other impoverished families begin new lives of health and hope. *Source: Heifer Project International.*

Whether to volunteer to just help out, or to attend for the purpose of gaining practical livestock and farming experience, volunteers should keep in mind that ranch and farming skills are not necessary.

Although the HEART Program and Heifer Project International have been featured, it is important to note there are many other formal and nonformal training centers that emphasize the teaching of appropriate technology and cross-cultural experiences. Some of these centers include: World Neighbors, World Hunger Relief, and the National FFA (Future Farmers of America) Association's Work Experience Abroad program. Other opportunities are presented in Appendix A.

Due to the uniqueness of Winrock International Institute for Agricultural Development located at Petit Jean Mountain near Morrilton, Arkansas, it was deemed essential to feature that institution as well.

Winrock International

Winrock International is a world leader in providing technical assistance and encouragement to agricultural development. Created in 1985, Winrock is making it easier for farmers in developing areas of the world to produce more food, fiber, and forests, thus

Figure 7–5 • In 1992, Heifer Project International is actively funding projects in 10 countries in Asia and the South Pacific that address the needs of small rural communities. Most of the programs in China and India involve heifer and goat dairy production. The projects in Thailand, the Philippines, and Cambodia are focused on draft animals providing people the needed power and manure for sufficient family food production. The Indonesia program has both components. *Source: Heifer Project International.*

improving their incomes and the quality of life for their families. This institute has a long-term commitment to reducing poverty and hunger in the world through sustainable agricultural and rural development.

Winrock helps people of developing areas in Asia, Africa, the Middle East, Latin America, the Caribbean, and the United States attain the following:

- strengthen their agricultural institutions.

Appropriate Agricultural Technologies **119**

Figure 7–6 • Heifer Project International is very active in India, working with Harijans and other people living below the poverty line. The goals are to increase home consumption of milk, to increase the family's income, and to reach many more families every year. *Source: Heifer Project International.*

- develop their human resources.
- design sustainable agricultural systems and strategies.
- improve policies for agricultural and rural development.

It provides services independently as well as in partnership with other public and private organizations. Winrock's activities are funded by grants, contracts, and contributions from public and private sources, and by income from its endowment.

Winrock's staff of more than 225 persons from over 20 countries have training and experience in planning and carrying out agricultural and rural development projects. In 1989, Winrock's staff provided technical assistance in 48 countries including the U.S. Staff members posted at 26 locations in 17 countries are responsible for implementing over 100 major projects, programs, and consultant assignments.

Winrock International is rooted in the philanthropic tradition of the Rockefeller family that, in the 1930s and early 1940s, helped to establish international technical assistance in agriculture in countries such as China and Mexico. This experience encouraged the Rockefeller and Ford foundations to join together to initiate what has become a system of international agricultural research centers. Research from these

Figure 7–7 • Takumi Izuno, Winrock specialist, assists farmers in a vegetable production field day at the Ayub Agricultural Research Institute, Faisalabad, Pakistan. *Source: Winrock International; photo by J. Cordell Hatch.*

centers led to the worldwide green revolution in wheat and rice production in the 1960s. Winrock International inherited that commitment and motivation. It was formed by the merger of three organizations with a common heritage of helping rural people of developing areas make better use of their agricultural resources. The three organizations that merged together included are the Agricultural Development Council (ADC), the International Agricultural Development Service (IADS), and the Winrock International Livestock Research and Training Center. Let us examine the latter more closely.

Helping people help themselves—that is the philosophy upon which the Winrock International Livestock Research and Training Center was established in 1975. Recognizing Rockefeller's deep personal commitment to agriculture and rural people, and his wish that a portion of his estate be used for charitable purposes, the trustees of the Winthrop Rockefeller Charitable Trust have supported Winrock International for more than a decade of growth.

Winrock International's central aims have always been to improve human nutrition and to provide income opportunities through the development of animal agriculture.

Figure 7–8 • Two forestry professors from India, M. H. Siddiqui (left) and B. B. Jadhav (right) studied tree crop production at Texas A & M University with Ronald Newton. Their training was managed by Winrock International under the Management Support Systems Project of USAID. *Source: Winrock International; credit by Texas A & M University.*

Implicit in those aims was the understanding that to achieve sustainable growth, the environment must not only be protected but also improved. The means for reaching those ends were never set in stone. Winrock has always been open to activities along the entire spectrum of agricultural production and distribution—from behind the farm gate to the halls of government.

Winrock projects fall into three program areas—national, international, and public-policy research. The work of the communications and conference center staffs span and support these program areas. There are strong links between project-classification lines. The results of a national project analysis of the grazing potential of forested lands might be used by policymakers as they set priorities for research and development. A communications project through which training videotapes are produced might be developed with the aid of international project personnel and might be used to train livestock producers in the Caribbean. The type of work Winrock scientists do falls into seven principal categories:

- conducting farming systems research to improve understanding and management of agricultural and human resources.

Figure 7–9 • Dupe Akintobi of Nigeria (right) is one of the first participants in Winrock's new program for African Women Leaders in Agriculture and the Environment. She is studying seed technology at Mississippi State University with Charles Vaughan. *Source: Winrock International; credit by Mississippi State University.*

- supporting and conducting training in research and development for host-country project leaders.
- aiding host governments in building research, education, and extension institutions to support agriculture in the future.
- providing technical assistance to host-country institutions in the development of their agricultural programs.
- analyzing for policymakers issues that have an impact on animal agriculture.
- producing materials on agriculture for training producers and fieldworkers and for informing both producers and the nonfarm public in the U.S. and abroad.
- conducting conferences, workshops, and seminars that directly contribute to the advancement of Winrock's objectives.

In less than fifteen years, Winrock has become an internationally trusted provider of information and expertise in animal agriculture. The institution has played a vital role in encouraging acceptance of the notion that investing in research, training, and development in animal agriculture is important, especially in Third World countries. Also, Winrock International has a publications catalog of brochures, videotapes, books, and conference proceedings that cover a wide range of information for livestock producers, scientists, and development institutions. The general focus is on animal agriculture; see Figures 7–7, 7–8, and 7–9.

The HEART program, Heifer Project International, and Winrock International welcome guests who wish to tour their facilities. The authors further emphasize that visiting and touring these programs is an excellent learning experience for U.S. and international students, as well as anyone who has an interest in agriculture either here in the U.S. or abroad.

Conclusion

It should be pointed out that technologies most appropriate in a given place and at a given time may well change, even from year to year. As environmental, political, economic, and social constraints are modified and as research and invention progress, the very appropriateness of a given technology may vary. Functional understanding and evaluation must be a concern of the scientist, the politician, the change-agent, and above all, the populace of each nation.

Discussion Questions

1. What meanings are generally associated with technologies of the various classifications?
2. Discuss the relationship between energy resource availability and agricultural technology.
3. How do you respond to the assertion that technologies imported from industrial countries sometimes run counter to cultural values and basic needs of the populace of other countries?
4. Why are so many technologies deemed appropriate for a developing nation classified as simple technologies?
5. What is the meaning of the term "ecodevelopment?"
6. Identify several technologies which have proven to be adequate for adoption and use by the developing nations.

CHAPTER 8
Dimensions and Implications of World Agricultural Trade

Source: *Heifer Project International.*

Objectives

After reading this chapter, you should be able to:

- discuss why it is important to the U.S. economy that trade-ties are established with other countries.
- know the meaning of comparative advantage.
- differentiate between tariffs and embargoes.
- explain which actions and policies restrict free trade.
- identify the developing countries which export agricultural products to the U.S.
- determine which foodstuffs are predominantly traded between nations.
- explain the role of agribusiness relative to a nation's economy.
- explain the role of farm managers relative to a competitive nation in a global market.

Terms to Know

comparative advantage
embargoes
tariffs
trade concessions
nontariff restrictions
transnational

transfer pricing	ISC	CWB
unilateral	EEC	CMEA
bilateral	CAP	protectionism
multilateral	subsidies	agribusiness
Public Law 480	levies	value-added
concessional	GATT	consumer-ready products

Introduction

History shows us that the exchange of goods among and between tribes, clans, and city-states has been a part of human activity for a very long time. Trade routes, both on land and at sea, were established early. However, the exchange of agricultural commodities, other than spices and goods having real or alleged medicinal properties, was limited. The major limitations were bulkiness and perishability. Today, with the rapid development of modern transportation, such physical limitations can no longer be recognized as major impediments. Nevertheless, among other remaining complex factors, a multiplicity of governmental rules and regulations has forced the necessity for treaties and agreements between trading partners. Yet, even with such an extensive and oftentimes complex base, the crux of the matter remains—how can each individual nation have a sufficient aggregate supply of high quality agricultural commodities to offer in the global market? Each nation's many primary or basic producers must have the knowledge, skill, and incentive to manage their own production units successfully. Farm and ag-business management remains the basic ingredient for each nation's success in the highly competitive arena of global agriculture trade.

Why Nations Trade with Each Other

All nations strive to attain a higher standard of living for their people. The limited number of nations that rank high on a worldwide scale of living-standard attainment certainly does include the U.S. This makes it a leader among the community of nations.

This wealth could not be maintained if the U.S. chose to isolate itself and not engage in a wide range of international trade activities. Exchange of commodities among nations often may become a controversial issue, but it has been demonstrated that the welfare of any nation is consistently improved by such interchanges. By its very nature, trade benefits all partners. This is fortunate, since no nation can be completely self-sufficient in providing for all of its needs. Therefore, a basic truth can be recognized in the assertion that "nations trade food with each other so that each country can be better off." A theory of long standing which applies to other commodities as well as foodstuffs is that of **comparative advantage.** This theory states that in a free market, those nations that take advantage of their resources, which when combined will allow them to put a product on the world's market at a cheaper price than other countries, will become major exporters of those commodities. A nation's comparative advantage is determined by four factors, all of which affect the cost per unit of output for domestic products; these include the following:

1. natural resource base.
2. location relative to markets.
3. production efficiency as measured by the ratio of inputs to outputs.
4. trade policy expressed in import or export tariffs and currency exchange rates.

A nation's natural resources and its location are generally considered to be somewhat permanent and therefore more difficult to manipulate for economic gain. However, the ingenuity of man can exercise a significant impact, especially on the other three factors.

As a consequence, the concept of comparative advantage has important implications for nations involved in international trade. Nations are not necessarily restricted to exporting only goods they produce best. In fact, they should produce goods they are relatively good at producing and allow other nations to do likewise. By doing so, all nations benefit and all nations, be they developed or developing, can identify some mix of commodities or goods advantageous to produce for export. Products that a given nation needs and wants, but is at a disadvantage to produce, almost invariably become goods that they import. Such demands for imports by any nation are associated with restrictions in growing the crop in sufficient quantity and quality to satisfy the needs or wants of its people. Import demands are often abetted by expanding populations, growing incomes, and changing tastes. Ideally speaking, as two countries trade with each other the needs of both are met.

However, much world trade is conducted under certain prevailing policies and regulations, and therefore can hardly be evaluated as being "free."

It should be recognized that every person is affected by the intricacies of trade carried out around the world. While most of us will never deal directly with the peoples of other countries, our lives are closely bound to theirs by an extensive network of trade relationships; see Figures 8–1 and 8–2. Almost everyone in this country has tasted cocoa picked in Nigeria, worn clothing stitched in Korea, or wrapped left-overs in aluminum mined in Jamaica. A labor dispute or bad weather in the most remote villages of these nations can affect the prices we pay for their products. On the other hand, the trade policies and business practices of our own nation can have a direct effect on the incomes of farm hands and factory workers in other countries.

The Importance of Food Trade

Food trade is important to both exporting and importing nations. The value of all agricultural products shipped from the country where they were grown to the country

Figure 8–1 • Braskalb Agropecuaria Brasileria, Ltd., an agribusiness consisting of storage silos in Brazil. *Source: DeKalb Plant Genetics.*

where they are consumed totals annually in the hundreds of billions of dollars. This represents a significant percentage of the value of total world exports of all commodities, including cars, computers, clothes, and other goods.

Because of the many different needs met by food trade, and the growth in world population, food trade has climbed rapidly in recent decades. In 1980, three times as much grain was traded between nations as in 1960. That equals 229 million tons of grain in 1980 compared to 76 million in 1960. It took 1,900 ships to carry that much grain in 1960; today it would take 5,725.

As the amount of food traded has grown, so has the share of it that is consumed in a different country than where it was grown. In 1960, 10 pounds out of 100 were consumed outside the country of origin; today that ratio is 15 pounds out of 100.

As was pointed out in Chapter 7, new technologies are needed in agriculture so that output can be expanded and farm profitability improved. It should be noted that progress does occur if proper public policies, investments, and other economic factors are attended to.

Simply stated, the series of changes proceeds in the following way:

1. As agricultural production rises and becomes more efficient, the agricultural

Figure 8–2 • A DeKalb drying plant in Argentina. *Source: DeKalb Plant Genetics.*

sector releases part of its labor force to industrial employment while still meeting the food needs of the nation.

2. As agricultural output increases, net farm incomes also rise, thus creating new levels of rural purchasing power and demand for additional industrial goods. Surplus income consumption that can be mobilized as savings to be invested in either industrial or further agricultural modernization is also created.

3. As agricultural output improves, relative food prices fall so that consumers are able to purchase food and other commodities at a lower price. A transfer of wealth from the agricultural sector to consumers in other sectors takes place where portions of salaries previously spent on food are now directed to higher value consumer-ready products or for nonfood purchases.

World Trade in Agricultural Products as Associated With Production Locales

Developed nations have a greater developed capacity to produce and process foodstuffs, and several of them do much more than is needed for domestic consumption. The factors contributing to a nation's production of excess foodstuffs which can be shared include the following:

- productive soils.
- favorable climate.
- available capital.
- monocultural practices.
- adequate allocation of natural resources to education and research.
- highly efficient technology.
- available instruction in management practices.

Kinds and Types of Agricultural Commodities Traded

All imaginable types of agricultural crops are traded among countries. Thirteen commodities make up the list of major agricultural commodities frequently entering world trade. Of these, eight can be considered as food crops. A listing of these commodities in descending rank as to world export value is given in Table 8–1. A handful of countries dominate world exports of food grains because of their large production of wheat or rice relative to domestic consumption levels.

It should be noted that most of the world's food supply is attained through the cultivation of relatively few plant species. Thomas, Curl, and Bennett (1982, pp. 23–24) pointed out that planted crop acreage in the world approximates 2.4 billion acres. About 1.7 billion acres, or three-fourths, are planted in cereal crops. Wheat, rice, and corn are grown on the majority of these acres.

Five countries of the world provide over half of the world's total cereal production, and include the U.S., China, the Commonwealth of Independent States, Western Europe, and Eastern Europe.

The U.S. is the world's leading grain producer, providing almost 20 percent of the world's total. China follows with about 16 percent, and the Commonwealth of Independent States with 12 percent. The U.S. is the leading producer of the coarse grains (corn, grain sorghum, oats, barley, rye, and millet), the Commonwealth of Independent States leads in wheat production, and China leads in rice production.

The dominance of North America as a global food supplier began in the 1940s. Exports expanded gradually during the next two decades; however, since the 1960s, exports more than doubled as a result of the dramatic population growth around the world. Today's leading U.S. agricultural export commodities include feed grain products, wheat products, soybeans, livestock products, horticultural products, cotton, tobacco, soybean cake and meal, rice, and soybean oil.

The U.S. is both dependent on world agricultural markets and is a significant trader in these markets. In recent years, more than 60 percent of all wheat, rice, and cotton

Table 8–1 •

Total Value Rank of World Gross Exports

Commodity Rank	Approximate Percentage Share from:	
	Developed Countries	Developing Countries
1. Wheat	90.0	10.0
2. Coarse Grains	85.0	15.0
3. Sugar	25.0	75.0
4. Oil Seeds	85.0	15.0
5. Coffee	4.0	96.0
6. Cotton	53.0	47.0
7. Tobacco	56.0	44.0
8. Rice	44.0	56.0
9. Cocoa & Cocoa products	24.0	76.0
10. Rubber	1.5	98.5
11. Citrus Fruits	66.0	34.0
12. Bananas	15.0	85.0
13. Tea	—	100.0

produced in the U.S. was exported. Furthermore, about 55 percent of soybean production and 30 percent of corn production was also shipped abroad.

Agricultural exports by developed nations to developing nations represent a significant and growing proportion of total agricultural exports. This trend has a logical base and indicates the ever-increasing importance of the developing nations to total world trade. There are two reasons why growth in farm exports to developing nations might exceed growth in farm exports to developed nations:

1. More than 50 percent of the world's population is located in the developing countries. Their very presence represents a need for food and food products, creating demand pressures for great amounts of food.
2. Most people in developing nations have low incomes, and people with low incomes spend a higher proportion of their income on food. While less than 20 percent of average income in the U.S. is spent on food, in India, Tanzania, and Niger, more than 60 percent of people's income goes for food purchases.

Wheat

Wheat is grown and consumed largely in countries located in temperate climates. Leading producers of wheat include the Commonwealth of Independent States at 20 percent, the U.S. at 15 percent, China at 13 percent, and India at 8 percent. Other countries with a lower production total, but important to world trade because of the high proportion of production exported, include Australia, Canada, and Argentina. The U.S. holds nearly half of the global wheat market. Canada has an extremely competitive wheat export marketing program including long-term agreements, extended payment options, credit guarantees, below market interest rates, and price discounts. Australia and France also have strong export marketing programs compared to the U.S.

It is interesting to note that while the U.S. produces only about two-thirds of wheat produced annually in the Commonwealth of Independent States, it is grown on one-third the total acreage used by that region.

Rice

The People's Republic of China is the world's largest producer of rice, with India being second; however, the U.S. exports more rice than either China or India. Indonesia ranks third in rice production. Some developing nations have recently been able to shift from being importing nations to being exporting nations. Thailand now exports a sizeable quantity of rice while Indonesia has greatly lessened the need for rice imports. In third place, behind the U.S. and Thailand, is Vietnam. During the mid-1980s, Vietnam was a net importer of rice, a recipient of food aid in order to avert famine. Today, net exports from Vietnam total some 1.5 million tons. This shift is largely due to the use of improved varieties, better cultural practices, and commercial fertilizers. Though the U.S. is a relatively minor producer of rice, accounting for less than 2 percent of world production, it is one of the world's largest rice exporters. U.S. exports account for one out of every four tons traded in world markets.

Feed Grains

As nations become more affluent, consumers often shift eating habits, desiring more meat and other animal products. This makes it profitable for farmers to shift to livestock production. Such shifts have brought about an increase of trade in feed grains. Currently, U.S. exports accounted for 74 percent of world trade in feed grains, with corn or maize constituting by far the largest component. However, in terms of total world coarse grain usage, U.S. exports account for only 10 percent of total world consumption; see Figure 8–3.

Oilseeds

Soybean protein accounts for about two-thirds of the world's production of protein meal. While the U.S. enjoyed a 94 percent share of the world market in 1970–71, it has currently dropped to 74 percent. This declining percentage is largely the result of sharply expanded soybean production in South America. Despite the declining market share, rapid growth in world demand has now pushed total U.S. soybean exports well above those of a decade earlier.

Figure 8–3 • A shelling plant in Chile. *Source: DeKalb Plant Genetics.*

Horticultural Products

Far more than most other agricultural commodities, horticultural products constitute a major phenomenon as evidenced in international agricultural trade. Trade developments occurring during the latter portion of the 20th century clearly reflect a climate and resource controlled exchange between temperate and tropical zones, therefore between developing and developed countries.

Classification of horticultural commodities entering the global trade markets include 1) fresh (largely tropical) fruits, 2) off-season fruits and vegetables, 3) bananas and pineapples, 4) citrus juices and similar products, 5) nuts, and 6) floricultural products.

In terms of world export value, bananas, as an agricultural commodity, rank 12th. Citrus fruits rank 11th, while cocoa products rank 9th. Banana exporting countries are grouped by areas 1) South America, 2) Central America, 3) Windward Islands, 4) Africa, and 5) Far East. Major importing countries include United States, European Economic Community members, Germany, France, Italy, and United Kingdom. Also, Japan and Scandinavia import significant quantities.

Citrus juices account for more than 90 percent of the world's output of citrus products. Concentrated orange is by far the most important item for international trade in citrus juices, accounting for 85 percent of exports of all types of concentrate and single strength juices. Brazil has become the largest processor and exporter of citrus products, with the U.S. second.

Cotton

As a major commodity in the area of agricultural raw materials, cotton continues to occupy an important position in world trade. Cotton's share in the global textile fibers market is maintained at around 50 percent in competition with jute, kenaf and other allied fibers. Expanding textile exporting countries, especially those in the Far East, are very dependent upon their ability to import raw cotton. Among agricultural commodity world gross exports, cotton is ranked 6th in terms of value. Developed countries have only a slight edge over developing countries in terms of value of cotton exports. Both production and volume of utilization are affected by size of storage, stocks, government programs, fluctuations of input, expenses and trends in cotton mill utilization. Major exporting developing nations and areas are Latin America, Africa, Near East, Far East and China. Exporting developed areas and nations include the U.S., Eastern Europe and a number of former communist states now in the Commonwealth of Independent States.

Animals and Animal Products

Leaders in beef exports are Argentina and Australia. Other nations exporting sizable quantities include New Zealand, France, Ireland and Uruguay. The Netherlands, Germany and Denmark also export substantial amounts of beef, while Mexico and certain Central American countries are beginning to make strides in increasing exports. African trade remains relatively insignificant largely due to production limits imposed by pests and disease. While the U.S. consistently outranks any other individual nation in beef and veal production, it is also consistent in being the major importer. Japan and the United Kingdom, along with some of the central European nations, also rank high in significant quantities of beef imported. As for pork, Denmark, long renowned for superior quality bacon and hams, is the chief exporter. The Netherlands, Belgium and Luxembourg follow, along with Poland, Yugoslavia and Hungary. The lone country of New Zealand accounts for over 50% of the international shipments of mutton and lamb, with Australia as the second most important exporter. Chief importers of lamb and mutton are the United Kingdom and Japan, followed by France, Greece, Canada and the U.S.

In response to the extensive admonition from nutritionists and dietitians, a significant number of consumers have changed eating habits to consume less red meats, more poultry and fish. Consequently, world trade has tended to decrease since it is more advantageous to produce poultry meat in or close to areas of consumption. Limited trading between neighboring countries does persist such as that between Canada and the U.S., China and Hong Kong or Thailand and Singapore.

Japan is the largest net importer of dairy products among developed market economies. New Zealand as a milk product exporter has significantly expanded sales to

Table 8–2 •

Countries That are the Best Prospects for U.S. Agricultural Exports*

Rank	Country	Rank	Country
1	Japan	11	Brazil
2	Taiwan	12	France
3	Canada	13	Pakistan
4	Hong Kong	14	Turkey
5	South Korea	15	Venezuela
6	Soviet Union**	16	Sweden
7	Italy	17	Mexico
8	Germany	18	Colombia
9	United Kingdom	19	Egypt
10	Spain		

*Source: "Foreign Agricultural Service." *Foreign Agriculture* Vol. 26, No. 2, Washington, D.C.: U.S. Department of Agriculture, February 1988.

**Now recognized as the Commonwealth of Independent States.

developing countries. National policies reducing subsidies to dairy farmers tend to be a major factor in shifting direction and dimensions of global trade in dairy products.

Because countries vary so greatly in their capacity to produce food crops, it would seem that a goal of total food self-sufficiency for every nation is neither economically nor developmentally desirable. However, for the foreseeable future most of the lesser developed nations must place high priority upon gaining self-reliance in acquiring adequate food for their populations.

As trade situations become more favorable due to the willingness of developed nations to accept commodities as imports from developing nations at prices comparable to those which poorer countries must pay for their imported foodstuffs, complete self-sufficiency may become an option for the developing country to consider. However, under prevailing conditions, world trade in foodstuffs will remain tremendously important for the foreseeable future. (Table 8–2)

The countries most involved in shipping agricultural products to the U.S. include the following:

- Indonesia
- Philippines
- Ivory Coast
- Guatemala
- Honduras

- Mexico
- Colombia
- Malaysia
- Ecuador
- Argentina
- El Salvador
- Brazil

Other leading agricultural trade partners include Canada, Australia, France, Netherlands, and New Zealand.

Actions and Policies Used by Nations Which Restrict Free Trade

Trade policies are a critical reality that must be faced as a result of the new order of economic interdependence among nations. Virtually every sector in the world economy is subject to the pressures and vacillations of world markets, and various agricultural policies also influence these international relationships. Even debt management problems of the developed and developing countries can be important to U.S. trade balances. The debt obligations facing nations like Mexico and Brazil mean that resources previously available for buying desired imports must be redirected toward debt payment.

Policy Examples Designed Primarily for Self-Protection

A number of policies and actions implemented by nations are designed primarily for self-protection. More often than not, the countries implementing such policies are the wealthy, more highly industrialized nations. These actions will now be discussed.

Embargoes. **Embargoes** are restrictions placed by an exporting nation upon sale of certain products to specified other countries. A classic example is the embargo placed upon wheat sales to the Soviet Union by the U.S. Almost without exception, embargoes are declared for political reasons.

Tariffs. **Tariffs** consist of duties or fees imposed by importing nations upon sales of products from exporting nations. Sometimes tariffs are imposed as a form of protection for workers and producers in the importing nation. Often involved is the phenomenon of much cheaper labor and other lower production costs occurring in the exporting nation.

Trade Concessions. Developed nations, particularly the U.S., engage in a practice of granting **trade concessions** to selected nations. Such action has political implications. The aspect of such action which may be detrimental to some of the world's hungry is that of discrimination. There is no escaping the conclusion that food often may be used as a weapon for imposition of political clout.

Nontariff Restrictions. **Nontariff restrictions** include quotas, packaging and labeling requirements, health and safety standards, and customs regulations. One document of GATT (General Agreement on Tariffs and Trade), lists over 200 nontariff restrictions.

Transfer Pricing. Quite often a **transnational** parent corporation is located in a developed country, with subsidiaries located in one or more developing countries. Through the operation of **transfer pricing,** the multinational corporation can adjust the price of commodities exchanged between the parent company and its subsidiary according to the rate of the country involved in the exchange.

Agreements Between Nations Which Determine World Trade

Many agreements governing the nature of policy in world trade are operating today. These can be classified by the following definitions:

Unilateral—relates to a contract or engagement imposed on one side or party.
Bilateral—affecting two sides or parties reciprocally.
Multilateral—participated in by more than two nations or parties.

Agreements or agencies affecting world agricultural trade will now be discussed.

The U.S. Agricultural Trade Development and Assistance Act, commonly known as **Public Law 480** or the Food for Peace Act, operates as a unilateral or bilateral agreement affecting trade. Under Title I of this act, food is distributed to needy countries in the form of concessional sales. The word **concessional** refers to favorable terms, including low interest, long payment periods, and occasional forgiveness of repayment requirements under which the sale was contracted. Under Title II provisions, grants in the form of commodities are provided to nonprofit U.S. relief agencies, friendly governments operating under bilateral agreements with the United States, and the World Food program, a joint understanding of the United Nations and the Food and Agricultural Organization (FAO).

These commodities are used to feed those people most susceptible to malnutrition, such as mothers and infants. In addition, they are used in food-for-work programs and for emergency relief. Totan (1982, p. 45) indicated that food-aid allocations for Titles I and II are decided by the Interagency Staff Committee **(ISC).** This committee is chaired by the Department of Agriculture and draws its members from the Departments of Treasury, State, Defense, Commerce, and the Office of Management and Budget. All of the aid allocation decisions by the ISC are made on the basis of consensus.

In brief, Public Law 480 intended to accomplish the following:

- expansion of international trade.
- development and expansion of markets for U.S. Farm Products.
- assisting people who are in need.
- promoting the economic development of friendly nations.
- furthering the foreign policy of the U.S.

The European Economic Community **(EEC)**, generally referred to as the European Common Market, is an example of an organization operating within multilateral dimensions.

This organization started in 1957 with six countries—France, West Germany, Italy, Belgium, Netherlands, and Luxembourg. In 1973 the United Kingdom, Denmark, and Ireland became full members. Greece, Turkey, Spain, and Portugal are now full members, and a number of African countries have become overseas associates.

Several trade restrictions among EEC members have been removed. A basic Common Agricultural Policy **(CAP)** has also been agreed upon. Trade among the member nations has increased substantially, and economic growth has been remarkable. The EEC also strives for self-sufficiency and increases in intracommunity trade through **subsidies** and **levies.** Particularly in the case of world marketing of grains, countries which are competing for markets but are outside the organization have been concerned over the EEC policy of granting heavy subsidies to their producers and thus being able to offer grains on the world market at comparatively low prices.

The General Agreement of Trade and Tariffs **(GATT),** consists of multilateral treaties between governments concerned with international trade. It has been in operation since January 4, 1948. Approximately 80 countries apply GATT in their international trading relationships. Members of GATT include nearly all the affluent free-market nations, some eastern European nations, and a more limited number of Third World nations. GATT provides a framework within which negotiations can be held for the reduction of tariffs and other barriers to trade, and a structure for embodying the results of such negotiations.

This organization discusses the import and export restrictions of the various nations and strives to remove those that restrict trade.

The Canadian Wheat Board **(CWB)** is the sole marketer of the grains for which it has authority—wheat, oats, and barley. All marketing of these grains is done by grain handling companies acting as agents of CWB and in accordance with the tariffs set by the board. Grains are delivered and regulated by producer quotas. Excess grain must be stored by the producer. Many other countries also have similar marketing boards for primary agricultural products.

The Japanese government's food agency controls most imported and domestic rice, wheat, and barley, and sets import levels and domestic price levels to carry out the government food policy.

The Council of Mutual Economic Assistance **(CMEA)**, founded in 1949, has operated as a union of socialist countries that was largely dominated by the Soviet Union.

The former Soviet Union had trade ties with 89 developing countries, and with 73 of them trade exchange was promoted by long-term trade agreements. The preferential trade regimes accorded by CMEA to developing countries in the mid-1960s in the form of free access for all products originating from these countries has accelerated trade turnover between the two groups. A series of long-term trade and economic agreements has assisted this process.

Now, however, there is uncertainty regarding the status of these long-term trade agreements, particularly since the democratic Commonwealth of Independent States has replaced the Soviet Union. A perfect example of this predicament is the future of Cuba.

National Benefits of World Agricultural Trade

Among the benefits to nations which may be derived from world trade are the following:

1. People are hungry because they are poor. Trade represents jobs and incomes, and provides opportunities for people to combat the cycle of hunger and poverty.
2. An increase in the importation of a highly competitive product into a country where the marginal cost in the country of export may result in a reduction of price often benefiting consumers.
3. Trade earnings pay for essential imports and help developing countries to qualify for international loans.
4. Importation of certain products by all countries tends to stimulate acquisition of new and useful skills in processing and manufacturing.
5. Under certain conditions, importing and exporting brought about by transnational or multinational corporations is beneficial, particularly for bringing in capital and providing employment. However, the claim is also sometimes made that the disadvantageous effects of such foreign investment in developing countries may more than offset advantages.

Aspects of World Trade That May Not Be Beneficial to Developing Nations

An imported product may sell at a price below that which must be received by producers in the importing nation for them to stay in business. Furthermore, an imbalance in trade between nations is a disadvantage to the country whose exports are restricted by the countries from which they must import. Such policies are often imposed by developed nations. Clausen (1983, p. 3), President of the World Bank, warned that **protectionism** is another persistent inefficiency in the developed and developing countries. The world economy cannot operate if all countries try to expand their exports but, at the same time, limit imports from other countries. Clausen (1983, p. 3) further stated that free trade fosters growth, and that cheaper imports of food, clothing, and other goods help in the fight against inflation. The alternative to protectionism is to assist people in declining industries, and to encourage the mobility of people and resources to industries that are internationally competitive. An adverse example of protectionism as granted to one sector of U.S. industry at the expense of another occurred when the U.S. lost over $550 million in wheat trade in 1983 in an attempt to satisfy domestic textile interests, which resulted in lower wheat prices, reduced export earnings, and a bigger burden for domestic farm programs.

Too often, the benefits of trade are concentrated among the relatively well-to-do minorities in developing countries, while the livelihood for the poor and hungry segments of the populace remains constricted. A government may seek popularity from citizens by negotiating cheap food aid imports while neglecting allocation of resources to the agricultural sector.

Most importantly, it should be recognized that imports and exports accomplished by transnational or multinational corporations are viewed by some as causing a situation which can place the developing nation at a relative disadvantage. It should be recognized that Third World countries have much to offer the firm seeking corporate investment, such as abundant resources, low cost labor, weak or nonexistent labor unions, and fewer environmental laws.

In many instances, tax concessions made by developing countries provide a fertile climate for corporate growth.

Agriculturally oriented transnationals sell grain and soybeans and market farm inputs such as tractors, pesticides, and fertilizers. The five largest grain companies control over 85 percent of world trade in grains. Transnationals exert much influence and market power over both farmers and consumers. A number of such services rendered by transnationals are judged as beneficial.

Flour mills, corn-compounding plants, and soybean-crushing installations often are as much symbols of status and prestige for a country as an airport and an international hotel; see Figures 8–4 and 8–5. But multinational companies rather than local governments are often in control of these facilities. They supply the capital, procure and install the technology, supervise the operation of the plants, control the procurement of the raw materials, and set up the local distribution systems for the dissemination of the processed products.

To the extent that transnational corporations enable countries to develop their own resources, provide employment for their populations, and obtain maximum benefit by having an equitable share in management and marketing policies, they must be assessed as beneficial. However, to the extent that they tend to exploit resources and employment, and avoid trade practices which might be of benefit to the host nation, there remain serious questions as to their social and economic impact.

Figure 8–4 • A seed plant in Nicaragua. *Source: DeKalb Plant Genetics.*

Agricultural World Trade and the United States

Historically, the U.S. farmer has demonstrated proficiency. In 1840, when our nation's population was 17.1 million, 4.4 million (25.7 percent) were farm workers. The average farm worker was supplying food for 3.7 persons (including himself) at home and 0.2 abroad. In 1969, with a population of 202.2 million, there were still only 4.6 million farm workers—little change in a period of 129 years—but the average farm worker was supplying food for 55.8 persons—42.2 at home and 13.6 abroad. The changes in productivity per worker were due principally to advanced mechanization and education directed toward more efficient management. Today, American consumers obtain their food for a lower percentage of their income than do consumers in any other country in the world.

Within the U.S., agricultural exports are not shared equally by all states. Illinois, Iowa, and California typically lead the list of states exporting agricultural products. The 11 important exporting states are in the Midwest, which reflects the basic importance of this group of agricultural crops in determining the level of U.S. exports.

These products represent a large part of the total volume passing through world markets. Export sales have important employment effects on the U.S. economy. It is

Figure 8–5 • Sorting and husking corn in Santa Helena, Argentina. *Source: DeKalb Plant Genetics.*

estimated that one million jobs in the U.S. depend on agricultural exports. Half of them are direct on-farm jobs and the other half are off-farm jobs related to agriculture. For each dollar generated by farm exports, two additional dollars are created in economic activity elsewhere in the U.S. economy.

WORLD FOOD TRADE, a publication of the World Food Institute at Iowa State University (1981, p. 6), stated that U.S. agricultural exports dominate world trade in a number of important products, and in some cases the degree of dominance is substantially greater now than a decade or two ago. Dominant shares of trade, however, do not indicate the U.S. has a monopoly in world markets. Competition in exports of major farm products remains keen, with domestic production in importing countries being part of that competition. However, dominant trade shares do indicate changes in U.S. supply, and demand conditions have a major impact on the quantity of grain available to world markets. Consequently, U.S. market conditions and government policies affect agriculture around the globe. Growing world trade also causes U.S. agriculture to be linked increasingly to world weather, economics, and political developments.

About 60 percent of our farm exports go to industrialized countries having relatively high incomes, including Japan, countries in Western Europe, Canada, and the Commonwealth of Independent States. A high proportion of the grain imported into these countries is used to feed livestock, although Japan does import substantial quantities of wheat for bread-making.

Smaller quantities of U.S. farm products are sales to middle and low income countries which have been experiencing rapid rates of income growth in recent years. This includes Taiwan, Brazil, and Mexico. Some of the petroleum-exporting countries have developed into major customers for U.S. farm products, industrial goods, and technology.

About one fourth of our exports go to countries at the lowest end of the income scale, where food supplies, even in good years, are close to a minimum. These include China, India, Pakistan, Bangladesh, and Indonesia.

The future of the U.S. agricultural and food industry is dependent to a large extent on the export market. The American people are not going to be consuming much more than they are already consuming. Therefore, in order to maintain a strong U.S. agricultural and food industry, it is crucial that we expand our agricultural export markets. Currently, Japan is projected to be the number one market for U.S. agricultural products; therefore, it is important to understand and appreciate the diet of the Japanese. This would also apply to other countries as well.

The U.S. Department of Agriculture's Foreign Agricultural Service analyzes and projects needs for selected U.S. agricultural products, and has projected that Japan will require the following products from the U.S.:

- high-value products (produced and packaged based on consumer demands).
- coarse grains and soybeans.
- forest products.
- bull and boar semen.
- frozen foods.

Japan's population is the world's seventh largest; Japan also has a limited area of land for producing agricultural products. When these two factors are considered in addition to the Japanese diet and favorable economic conditions, Japan can be considered an excellent future market for U.S. agricultural products. Contrastingly, Taiwan will require the following products from the U.S.:

- fruits and nuts.
- beef and poultry.
- wine and beer.
- french fries.
- wood products.
- feed grains.
- soybeans.

Canada will require the following:

- fruits and nuts.
- vegetables.
- grocery items.
- cotton.
- fast foods.

Hong Kong will require the following:

- fruits and vegetables.
- fur skins.
- poultry and beef.
- grocery items.
- bread and pastas.
- forest products.

South Korea will require the following:

- forest products.
- cattle hides.
- leather.
- fruits and nuts.
- juices.
- turkey and poultry.

The Need for Agribusiness Involvement

Agriculture is no longer confined to farming (production agriculture), particularly in the developed nations. Food and fiber products are delivered to consumers in these countries at a cost that requires a greatly reduced portion of individual income. Of course, this has only just begun to take place in most developing nations. Within the food and fiber systems of the more fully developed countries, input industries (e.g., agricultural chemicals, fertilizers, machinery, supplies, farm credit, and conservation), and output industries (e.g., processing, marketing, and transportation) contribute to producing both on-farm and off-farm jobs and profits. The output industries in particular have the ability to collect and process food and fiber, and to move it to areas where it is consumed.

In communities where agribusiness firms have become established, the productivity of farmers has increased because the farmer's knowledge and understanding of production, marketing, and processing has been increased. The U.S. and other developed countries not only have a more efficient food and fiber system, but one which is accompanied by a responsible and totally integrated agribusiness complex as well. The process of implementing a totally integrated food and fiber system is often best accomplished through continued involvement of those in the private sector. Individual and business initiatives provide an invaluable source of expertise for implementation. Government programs generally tend to address government-related problems rather than problems which are more immediately confronting farmers and agribusinesses; therefore, any attempt toward achieving the goal of increasing agricultural productivity, employment, and profits must have input from private farm and agribusiness interests.

Robert Spitzer, former U.S. Coordinator of the Food for Peace Program, stated that where agribusinesses are included, as in Taiwan, excellent food systems have been created (1981, p. 207). Spitzer implies that agribusiness involvement is essential if a country is going to be successful in feeding its people.

The U.S., Japan, and many other developed nations have strongly encouraged development of an agribusiness complex in the developing countries. The U.S. has a number of large multinational agribusinesses that are involved in the development of agribusiness intiatives in many developing countries. These agribusinesses provide technical expertise in the assessment of resources, management models and sytems, and employment. When agribusiness is integrated into the food and fiber system of a developing country, the farmer becomes more productive because additional assistance and incentives are provided. Employment created by agribusiness development in the local or rural community slows the migration of farmers and farm workers to the larger cities.

More U.S. agribusiness firms are seeking ways to expand into the international arena. However, many agribusiness firms have been reluctant to do so for the following reasons:

1. they are preoccupied with their own local operations and this month's profits.
2. some have inferiority complexes on world matters.
3. some lack understanding of governmental involvement.
4. some simply are not interested.

In the U.S., attempts are being made to internationalize our public school and post-secondary curriculum in order to increase understanding of international agribusiness involvement.

As U.S. agribusinesses strive to expand in the international arena, more people with international agribusiness experience will be required. Currently there is a shortage of trained professionals with such expertise.

Another example of the importance of agribusiness is Bulgaria's request for assistance from U.S. institutions to assist in the development of a university in Sofia. Bulgaria has a network of highly qualified scientists in 32 agricultural experiment stations which are located throughout the country; however, extension and agribusiness programs do not exist. It is anticipated that the scientists will form the core of the teaching faculty and they will receive assistance in providing technical training in areas of business and agribusiness. Concurrently, the government of Albania is seeking the same type of assistance.

Indonesia and Poland are also seeking agribusiness assistance. Indonesian agribusiness firms (primarily plantation-based firms) are in the process of becoming vertically integrated for value-added products. The commodities include palm oil, rubber, cocoa, coffee, tea, fish, hybrid seeds, quinine, and spices. Polish agribusiness leaders are interested in workshops designed to assist them in understanding business strategy alternatives during the transition to a market-oriented economy.

World Trading Patterns as Affected by Quality Consumer Readiness and Value-Added Features

Value-added, as applied to agriculture, is the difference between raw or bulk commodity input costs and the price of finished outputs. Examples include processed or refined cereal grains, processed meats and dairy products, vegetable oils, and other canned, frozen, and microwaveable food items. Unprocessed but relatively expensive **consumer-ready products** such as eggs, nuts, and fresh fruits and vegetables are categorized as high-value foodstuffs. It should not be overlooked that the maintenance and enhancement of quality standards also contributes to higher values.

Consumer demand or preference for more highly refined, processed, or expensive consumer-ready products is largely a function of the standard of living achieved. Rising income levels often contribute substantially to changes in consumer preferences or demand for such products.

Less developed nations, lacking necessary processing equipment and resources that might enhance the attractiveness and desirability of consumer-ready products, are at considerable disadvantage because their exports of agricultural commodities must be confined to products in the raw or bulk stages.

In competition between developed nations for export markets, the margin provided by application of value-added dimensions may often provide an advantage. Economists point out that implementing such programs enlarges the scope of economic activity, in essence providing an export market for domestic goods and services required to assemble, process, and distribute value-added and high-value consumer-ready products.

Staying Competitive in the Global Market*

The U.S. has generated an agricultural trade surplus every year since 1960. Over 40 percent of the total farm output was exported in 1980, and reflects the internationaliza-

---

*Adapted from the U.S. Department of Agriculture.

tion of U.S. agriculture. Approximately 20 percent of farm volume is exported, which accounts for 25 percent of farm cash receipts.

In the 1980s, U.S. agriculture experienced a significant period of adjustment. The stagnation of the world grain market led to a buildup of U.S. and world stocks and lower prices. Domestic fruit and vegetable producers also faced stiffer competition from lower-cost imported products, while important crops such as wheat, corn, and soybeans experienced more competition from subsidized production and exports in the European Community and South America. Concurrently, U.S. farm policies were in conflict, resulting in higher loan rates.

In contrast, world grain and oilseed trade doubled during the 1970's, with the U.S. snaring three-fourths of the increase. U.S. farmers increased production rapidly by farming land that had been idled during the land diversion programs of the 1960's, and increasing yields by using more inputs and using labor and capital more efficiently. Farmers borrowed heavily during the 1970's for investments to further increase capacity.

Recent declines in export volume and world market share, growing surpluses, and increasing food inports have caused concern about the future of U.S. agriculture. Major adjustments have affected not only the farm, but agribusiness and rural communities as well, and this process is expected to continue.

International Competition

International competition takes place in a dynamic framework. Production capacity, infrastructure, currency exchange rates, export/import restrictions, credit policies, shipping rates and regulations, international policies, and other factors affect a country's economic position relative to that of competitors.

To be competitive, a country has to secure a profitable share of a specific market in spite of efforts of other nations to secure the same market. Production costs, infrastructure, and government policy are important factors affecting international competition.

One of the major factors in a country's ability to compete internationally is that of production costs. These costs are affected by the prices of seed, fertilizer, land, labor, capital, and technology, as well as by the conditions under which crops are grown. As agricultural productivity increases, input costs per unit decline.

However, being the lowest cost producer does not always ensure competitiveness in today's market. Infrastructure—including transportation, communications, electricity, roads, and storage facilities—is critical in determining the cost at which a country can produce and deliver a product to the international market.

Achieving Goals and Objectives in an Internationally Competitive Market

Farm managers will have to develop better financial management and marketing skills to stay competitive in the world market. Farm managers with the ability to adjust to the changing global market will be the most successful.

Declining government intervention in agriculture will result in more flexibility in farm decision making. Farm managers will learn that gains in returns for improved marketing and management skills will far exceed those for increased production efficiency. Little capital investment is required for this as compared to the purchase of new equipment, and requires minimal cost.

Farm managers who adopt new information technology quickly can expect high dividends. With low-cost electronic information transfer, costs in new computer hardware and software can be recovered in a short time. The use of computer-based decision aids will become especially important to farm managers.

Internationally, competition will remain strong, especially in Argentina, Australia, and Canada, where agricultural production operates with little or no subsidy. In the European Community where agriculture has been supported above the world level, declining grain output and accelerated consolidation at the farm level will occur. U.S. farm managers will be well positioned to respond to this changing world environment.

Management Functions of Producers Basic to World Trade

All nations aspiring to become exporters of selected agricultural commodities must develop the ability of their producers to efficiently produce a sufficient quantity of these commodities.

Although a tremendous gulf is seen to exist between producers in the developed and developing worlds, basic management functions are shared by both. A major task facing those who promote growth in developing countries is that of assisting producers to understand, adapt, and implement proven management practices. Textbooks stress management functions as a set of goals and objectives for a business, developed and understood by the owner, management, and labor. Expectations concerning annual earnings, production, maintenance of farm buildings and grounds, tradeoffs between capital appreciation and earnings, long-term growth, and achievements must be established.

Figure 8–6 suggests the five basic management functions or activities required to achieve the goals and objectives of business. These activities will be discussed in detail.

Planning. While all five basic functions are important, planning is critical because a good plan involves all the other functions. Planning requires the setting of daily priorities and schedules, recognizing problem areas and looking for alternative solutions, making an annual financial plan and cash flow statement, looking at alternative cropping plans for the year, establishing the overall enterprise for the business, and developing the business.

Organizing. Organizing requires establishing an internal structure of the roles and activities required to meet goals. It also means deciding the functions of each position, and establishing the work routines and standard operating procedures for each production enterprise.

Figure 8–6 • Functions of farm management. *Source: United States Department of Agriculture.*

Staffing. Staffing is crucial to both small or part-time businesses and to larger ones. Often, the need to determine how to get all the tasks completed on time is even more critical because there is little flexibility in the labor supply.

Directing. Directing is closely related to staffing. The smaller the business, the more the two are interwoven.

Controlling. Control is the part of business management that determines which new methods are needed to turn out positive results when an investment decision is proven to be less profitable than planned. Control requires keeping track of expenses and income. Controlling includes the monitoring of records and accounts for the operations, comparing rates of production and levels of performance against established goals, and monitoring production processes and making changes as necessary.

Total farm management can be viewed as the systems approach to running a farm. It is a complete way of looking at the operation. While efficient farm management has always been important, it is even more critical in these rapidly changing times.

In brief, farm management can be defined as the coordination and supervision required to increase long-term profits or achieve other goals. It can be viewed as a combination of production management, business management, financial management, marketing management, and personnel management.

Conclusion

Planners and leaders throughout the world must understand the basic importance of agriculture to overall economic development. Too often, agriculture is viewed as a tradition bound sector with the sole mission of producing food. Agriculture is not seen in the broader context as the principal source from which overall development can emanate.

Rising agricultural production sets off changes throughout a country's economy that results in farmers producing a food surplus that can be marketed for cash or exchanged for other goods and services. Since people first choose to improve their own diets, much of the new income is spent on food, which promotes agricultural progress and improves nutritional levels.

A high concentration of people in the agricultural sector is a fundamental characteristic that distinquishes low-income nations from industrial nations. Among the lower-income nations, Chad has 85 percent of its population working in agriculture, Nepal has 93 percent, Niger has 91 percent, and Bangladesh has 74 percent. By contrast, the U.S. and England have 2 percent of their populations employed in agriculture, West Germany has 4 percent, and Japan has 12 percent.

Trends in future world trade will be dominated by three characteristics: a growing magnitude of trade, an increasing dependence of the world's poorest regions on trade to improve the diets of their rapidly growing populations, and an increasing importance of a small number of developed nations as the chief source of trade in primary food commodities.

The developing countries represent a tremendous potential market for U.S. goods and services, provided that the developing countries have the financial resources with which to purchase U.S. products. By the year 2000, there will be an additional 1.8 billion people who could be potential consumers for U.S. products, and most of those people will be living in a developing country. The demand for U.S. products by developed nations will continue to be rather stable; however, the demand for U.S. products by the developing nations will offer the greatest trade potential.

Discussion Questions

1. Why is the export of agricultural products so important to the economy of the United States?
2. Explain why the theory of comparative advantage must be considered in light of free-market conditions. How does protectionism affect free-market conditions?

3 Identify some actions or policies used by nations which tend to restrict free trade.
4 Explain why both developed and developing countries are effected by the vagueness and intricacies of agricultural trade.
5 What factors can contribute to a nation's production of excess foodstuffs which then can be exported?
6 List the developing countries which frequently ship agricultural products to the U.S.
7 Which types of food or other agricultural products are most often traded between nations?
8 What has Public Law 480 intended to accomplish?
9 What do developing countries have to offer U.S. firms who are seeking corporate investment?
10 What changes are set off within a nation's economy when farmers produce a food surplus?
11 What are the five basic farm management functions or activities used to achieve the goals and objectives of the farm business?
12 What must a nation do in order to remain competitive in a global market?
13 Explain what is meant by value-added products. List some examples of value-added products and explain how these products might enhance economic activity in a developing country.
14 What is the difference between value-added products and consumer-ready products?

CHAPTER 9

Political Aspects of World Food Production

Source: United Nations Development Program; Photo by P.S. Sudhakazan.

Objectives

After reading this chapter, you should be able to:
- understand why political strategies carried out by nations can be major determinants of the nutritional well-being of its populace.
- distinguish between politics and political.
- distinguish between corporate, group, state, and family farms.
- understand the importance of necessary credit and other services to farmers.
- understand the basic assumptions upon which U.S. policy is developed.

Terms to Know

political
politics
agricultural and food
 policy
land tenure
corporate farms
state farms
group farms
family farms
support system
rural village centers

The Effects of Politics on World Food Production

Governments commonly intervene in their food and agriculture sectors because of the fundamental importance of production and distribution of food. As agriculture has become more industrialized and as trade as grown in proportion to world consumption, the importance of government has become greater. Today, food and agriculture are affected as much by macroeconomic, trade, and development policies as by policies directed specifically to food and agriculture.

Political and Politics Defined

Political, for the purposes of this text, can be defined as "of, relating to, or concerned with the making, implementation, and enforcement of governmental policy." **Politics** refers to particular strategies used by a body to win support, approval, or power for application of an underlying ideology for overcoming constraints experienced within a society.

Political strategies carried out by each nation can be major determinants of the nutritional well-being of its populace. Unquestionably, food crises are not only a matter of food shortages, inadequate nutrition, economic conditions, and population policies, but also a matter of politics, both national and international. Politics is one of the underlying causes of the current agricultural production and food distribution problems. One can divide the issue into two—the internal problems and policies of nations, and the interdependence of nations for food supplies and agricultural equipment, though the two influence each other.

The professional politician has slowed progress in reaching sustained agricultural development objectives, even though the professional politician is seldom mentioned in discussion on global food production. They must adopt, finance, advocate, and defend the policies, programs, and projects recommended by scientific and technical professionals.

The professional politicians see the policies to be formulated, choices to be delineated, and issues to be resolved from quite different perspectives than do agricultural professionals.

Decisions which are made by the government or "power elite" of each nation have tremendous effect upon the physical and mental well-being of the nation's citizenry. Malnutrition and famine can ultimately be charged against political realities not only in the affected country, but also in all nations. Too often the foreign policies of one nation may be significant for the welfare and ultimate destiny of millions in other lands. The effects of foreign and domestic policies implemented by a ruling government and aimed toward gaining advantage in world relationships may prove to be just as crucial to their own populace.

Agricultural and Food Policy

In the U.S., public policy directly affects the economic well-being of farmers. Decisions made by government agencies often impact farm income as much as do management and marketing decisions made on the farm.

Agricultural and food policy includes land-use and conservation policies, marketing policies for both input and product markets, foreign trade policies, nutrition and food distribution policies, and commodity price and income support policies. The purpose of public policy research and educational programs is to provide information on options and their impacts that legislators and farmers can use to make decisions.

An Example of Unprecedented Political and Economic Change of Agricultural Policy:

Poland. According to Penn (1989), the events in Poland today are truly historic, unthinkable only a few years ago, and occurring at an unimaginable pace. Politics and economics are intertwined, and food and agriculture are playing a central role. Political stability, supported by patience of the Polish people in enduring further hardship, is critical for a successful transition to a market-oriented economy.

A non-communist government, the first in over 40 years, was installed and pledged to change the centrally-planned socialist system over to capitalism without delay.

The challenge of achieving political reform, developing markets, and an availability of consumer goods is appearing first in the area of food and agriculture. According to Penn (1989), this is not unusual in Polish politics. In fact, scarcity of food and related high prices precipitated the collapse of three previous governments in 1956, 1970, and 1980. Food is no less central today, and is the element most critical to the success of the economic transformation now beginning.

Food occupies a prominent political role in Poland for several reasons:

- the severe shortages that twice in this century were experienced by Polish people during wartime conditions are fundamental.
- as in many low-performance economies, food is practically a preoccupation in everyday life.
- food purchases require a very high proportion of disposable income, estimated to be as high as 65 percent.

Penn (1989) concluded by indicating that Poland's economic system is barely functional at best. Market institutions for guiding production, distribution, and consumption in many cases simply do not exist. The pressures for their development and for them to deliver increasing supplies of consumption goods at reasonable prices are enormous. Currently, food supplies are uneven, prices are extremely high, and the hardships on many Polish people are immense.

After decades of centralized planning, Poland is attempting to build an infrastructure for a successful market economy. Polish political leaders have indicated a need for knowledgeable people who can successfully manage private enterprises and understand sophisticated market economies. Particularly, Polish political leaders are interested in the following:

1. creation of new, commercially viable private agricultural cooperatives and other private agribusinesses.
2. enhancement of the commercial performance of existing private agricultural cooperatives and other private agribusinesses, principally in processing and marketing.
3. creation of agricultural credit institutions to provide credit and banking services to farmers and private agribusiness.
4. creation of new and productive relationships between agricultural extension services and private farmers.
5. privatization of state agricultural enterprises that will provide services to private farmers and their organizations.

It is believed the agricultural sector will play a key role in the transition from a planned to a free-market economy.

Bulgaria. The U.S. Agency for International Development (USAID) has provided a $10 million grant to assist Bulgaria in developing a free-market system.

The grant will support Bulgaria's efforts to return collectivized farms to their former owners by helping to pay part of the costs of 1,500 local land councils that have been established to carry out the privatization program.

The funds will be used to import goods from the U.S. for sale in Bulgaria. The local currencies generated will go towards the cost of processing land claims.

Bulgaria has enacted privatization laws and established procedures to select key personnel to manage the process.

The grant, which was signed by Vice-President Dan Quayle, marks the first time the U.S. has provided bilateral assistance to that country; refer to Figure 9–1.

USAID is also working with the World Bank and other donors to support additional incentives for Bulgaria to initiate and carry out agricultural policy reforms.

Policies Regarding Enforcement of Military and Police Powers

Regardless of one's belief in the "rightness" or "wrongness" of political actions of any nation, military action often results in malnourished or starving people. Hundreds of thousands fleeing across borders to seek refuge from either invading enemies or from civil strife are irrefutable evidence of the horrendously effective power of armed conflict. Likewise, the imposition of despotic and inhumane regulations by government

Figure 9–1 • U.S. Vice-President Dan Quayle participates in a signing ceremony (July 29, 1991) marking the first U.S. bilateral assistance agreement with Bulgaria. *Source: United States Agency for International Development; photo by Clyde McNair.*

upon its citizenry inevitably results in a lack of nourishing foods for the people remaining in the country, and those seeking asylum in refugee camps elsewhere.

Starvation is menacing Africa once again, the all too familiar crisis brought back by the deadly combination of lack of rain and the chaos of war. According to USAID surveys, drought and civil strife have placed an estimated 20 million to 35 million people in sub-Saharan countries at the risk of starvation (USAID HIGHLIGHTS, Summer, 1991).

In Angola, Liberia, Sudan, and Ethiopia, government policies have tended to favor urban consumers over rural producers, thereby discouraging local food production and contributing to food shortages. When these policies are coupled with civil strife, farmers are unable or unwilling to plant. One hope for Ethiopia, now that the Socialist regime is being dismantled, is that the market will be allowed to determine agricultural prices, which will serve as an incentive to greater local production. Angola is facing its own food emergency that may affect almost 2 million people. The emergency stems from the cumulative effects of four years of drought and 16 years of civil war and economic stagnation. Drought in central and southern Angola has reduced agricultural production by 75 percent in the region. More than 1 million of Angola's rural population left their homes and farms to escape prolonged fighting. In Liberia, more than half the population has been affected by 18 months of particularly bloody civil war. Within Liberia, 1.2 million have been displaced, and more than 750,000 have fled to neighboring Guinea, Cote d'Ivoire, Sierra Leone, Ghana, Nigeria, and Mali.

Approximately 15 years of constant civil strife, nationwide food shortages, and economic stagnation have taken their toll similarly on Mozambique. Much of the country's productive assets and infrastructure have been destroyed. Mozambique has almost 2 million internally displaced people, and more than 2 million more have sought refuge in surrounding countries. Another 2.5 million living in rural areas have been unable to grow enough food because of chronic drought and frequent fighting.

Except for Sudan, there are signs that the civil wars that have raged for so long in these countries may be drawing to a close. Peace would mean increased agricultural productivity and also greater ability to deal with crop failures when they do occur; see Figure 9–2.

Policies Regarding Food Acquisition and Distribution

Cheap Food for Consumers

One of the most discouraging policies for farmers is one in which their government concentrates, often for political reasons, on providing cheap food for urban consumers. There are documented cases in which a government, seeking approval of the masses of its citizenry, has acquired concessionary grain imports and sold them at prices much lower than the cost of production for its own farmers. In the long run, such actions are extremely counterproductive; they greatly reduce incentive for domestic producers, and tend to encourage massive migration to the cities.

Trade and Food Distribution Policies

As alluded to in Chapter 8, policies regulating trade between nations not only affect the acquisition of foodstuffs through imports, but also govern the quantity and quality of domestically produced foods distributed within the developing nations. Government policies function oppressively when they exert undue pressures for production of crops to be exported, at the expense of production of nutritious foods greatly needed for domestic consumption.

Figure 9–2 • As part of a series of June 1991 press events on the famine and relief efforts in the Horn of Africa, Assistant Administrator for Africa, Scott Spangler, briefs foreign journalists at the U.S. Information Agency Foreign Press Center in Washington, D.C. In his presentation, Spangler emphasized the need for government reforms to ease the famine situation: "When we look at the situation in Africa, it is clear that droughts are natural occurrences, but famine is man-made. Governments must take responsibility and establish the political and economic framework for increases in production and more equal access to food by all members of the society." *Source: United States Agency for International Development; photo by Clyde McNair.*

Policies Regarding Agricultural Production Incentives and Support

The prime goal of agricultural policy in any nation would be to provide the nation's people with an adequate, sustained supply of high quality, nutritious food at low cost. However, care must be exercised to avoid such simple solutions as price fixing of foodstuffs at such low levels that it may possibly function as a discouragement to the nation's farmers. Not only do farmers need financial incentives to grow the nation's requirement of food crops, but they also must feel that they constitute an important resource. Unfortunately, farmers are too often ignored in the making of political decisions. In many countries the farmer has no voice in making the decisions that affect his profession. Decisions as to planting materials, methods of operation, type of lending institution, and the system of **land tenure** considered suitable for improved agricultural production are made at national levels and handed down at provincial headquarters for implementation. This results in a misunderstanding of government motives, and the mislabelling of cautious farmers as conservatives who prefer traditional ways to modern, scientific innovations designed to improve their output.

Land Ownership and Tenure

Just as countries can be classified according to political-economic categories, farm enterprise systems can also be classified largely with regard to organization, land use, and tenure. An understanding of the various farming systems is necessary for the most efficient application of agricultural support policies. Farm enterprise systems are usually grouped into the following four basic categories:

1. **Corporate farms,** are large-scale vertically-integrated production units. The entire system of production, harvesting, handling, and marketing is usually controlled by a single organization.
2. **State farms,** also quite large, are operated by governments. Managers are directly responsible to a bureaucratic unit, and employees generally work for wages.
3. **Group farms,** incorporating some aspects of cooperatives, are operated jointly by members. Income is generally allocated according to an individual's share.
4. **Family farms,** operated by families, vary greatly in size and production rates. Family income depends upon productivity, and incentives to improve productivity are strong.

While the majority of producing land units in most nations consist of very small family farms, group and state farms can also constitute a sizable portion. Various combinations are also found in some areas. In Africa, land is often owned by a group and not by individuals. The group may be a village, a clan, or a family. By virtue of membership in the group, an individual has access to the use of the land.

Recent research has shown that, compared to large farms, small farms (when farm size refers to the operating unit, not the owned unit) tend to be more efficient operations. It would seem that government policies favoring land reform should prove beneficial not only to individual "campesinos" or peasant farmers, but to the nation as a whole because a more plentiful supply of food may result. The objective of most development planners in the Western world is to achieve a mix of farm enterprises throughout the nation and within a commercial family farm economy.

Policies Regarding Use and Allocation of Resources

Each nation has varying amounts of resources, both developed and undeveloped. Resources may be classified as natural, renewable, unrenewable, and human. Judicious use of natural resources is of primary importance for the progress of each nation. No less important is the dissemination of knowledge and acquisition of skills through the human resource of educators, professsionals, laborers, and entrepreneurial managers. The percentage of budget allocated to education and extension, as well as provisions for upgrading and retraining workers, may well be determining factors in the progress of a nation.

Government Efforts Toward Establishment and Maintenance of Support Systems

A major requirement for progress in all nations is that support systems include the provision of necessary credit to local farmers. Also important is the willingness of government to allocate sufficient resources both to teaching (adult education and extension) and to agricultural research. In the case of credit and informational services, a means of economically and effectively dispersing support and reaching the rural dwellers must be found. It is the responsibility of governments to provide education and health services of the same quality to people in the countryside as to those residing in the urban sectors. Failure to do this may limit the incentives of farmers to improve production and cause their children to migrate in large numbers to cities. Government also has responsibility for providing research facilities and organizing an efficient extension service.

The growth of the economy is dependent on the simultaneous implementation, maintenance, and expansion of supportive services. The **support system** plays as crucial a role as scientific innovations in the effort to increase farm output. To be able to purchase new and improved inputs, farmers need access to cheap sources of credit; they will also want assurance that these inputs are conveniently located and are available at reasonable prices. Fertilizers will not be used if it involves travelling long distances to supply depots and paying high prices. If there is no facility for storing the bumper harvests, or a ready market does not exist for the products of new technology, farmers will stick to their traditional subsistence crops instead of taking what appears to them to be avoidable risks.

The organization of an efficient system of agricultural services involves considerable financial outlays and a corps of highly skilled managers to operate them. The absence of these factors has hampered many developing countries.

The Need for Rural Village Centers

Too often farmers and rural dwellers in some countries are passed by when efforts are made to provide support services through government agencies. Villages are remote and often considered inaccessible. Roads remain primitive, and consequently adequate transportation is lacking.

Governments need to encourage establishment of concentrated **rural village centers** of 2,000 to 20,000 people. Such villages could then provide the following elements of the support system:

- banks.
- markets for farm products.
- farm supplies and equipment.
- machinery maintenance and repair services.
- schools, health and recreational services.

Government support of road building to link villages to urban areas would be of maximum value.

U.S. Policies Regarding Assisting Developing Nations

U.S. policy changes somewhat due to political changes (i.e., party persuasions, lack of popular support for foreign aid, or economic depression). However, over the long run, most policy decisions are based on the following assumptions.

1. Developing countries should constantly reaffirm their belief that the primary responsibility for insuring their development rests with themselves. They should intensify their efforts in agricultural development and food production, including the eradication of hunger and malnutrition.
2. New technology will increase production more than changes in land tenure, land reform, or produce markets.
3. Under free-market conditions, food self-sufficiency in all nations is neither a desirable nor an efficient use of resources. Those countries able to produce a food product most economically should specialize in that product, and sell the surplus to nations needing the product. Certain developing countries must improve their overall food production because they do not have the currency exchange to purchase needed food. The developing countries must be willing to devote sufficient resources to expand food production. A national policy must be implemented by each nation to give agricultural production high priority.

4. Governments must work together to achieve higher food production and to implement means for a more efficient distribution of food. Governments should formulate appropriate food and nutrition policies integrated into socioeconomic and agricultural development plans based on adequate knowledge of resources. Foremost, the responsibility of any government is to its own people.
5. Most countries will have some technical assistance they can provide to other nations. Many agricultural colleges and governments can secure the services of trained technicians to help others. Private investment made in another country may often carry with it technical skills and advanced equipment.
6. Each nation should evaluate its food resources to determine the amount available for sale or aid. The amount of food aid given by any nation will depend on the availability of supplies and the relative costs of such aid. The increased cost of petroleum increases the cost of food production.

A succinct five-point statement of desirable policy for the U.S. was expressed by Alexander Haig, Jr., former U.S. Secretary of State, and is as follows:

1. In the formulation of economic policy, in the allocation of our resources, in decisions on international economic issues, a major determinant will be the need to protect and advance our security.
2. We shall continue to work with other countries to maintain an open international economic system. This will include efforts to engage the U.S. private sector more fully in the economic development process.
3. The U.S. will not forsake its traditional assistance to the needy of this world—the undernourished, the sick, the displaced, and the homeless.
4. There will be neither abrupt nor radical redirection of our international economic policies. Where necessary, policy will be changed in an evolutionary fashion with minimal disruption and uncertainty.
5. The U.S. will not abandon institutions and agreements devoted to global economic and political stability. The U.S. will continue to bear a fair share of the cost to maintain and operate international organizations.

Conclusion*

There is much evidence worldwide that farmers are economically rational. Farmers invest money, plant crops, and adopt new techniques when they believe such actions are in their best interest. The marketplace provides the information farmers need, especially

---

*Adapted from the National Association of State Universities and Land Grant Colleges.

with respect to pricing. Information must be accurate for good decision making. This occurs when prices accurately reflect the value of scarce products, services, and resources.

Farmers are dependent on support systems to assist in production and marketing, and need assistance in learning how to gather, process, and utilize new information. They also need government to monitor and protect them.

These interventions and public policies represent a set of government actions that affect incentives for farmers. A proper framework of public policies is essential for production incentives.

Worldwide, farmers need to be as cognizant on financial management, farm management, and public policy as they are on biological and technical matters. Given the importance of management expertise in agricultural production, and the rapid changes likely to occur in the decades ahead, research and educational programs in farm and financial management and public policy are essential.

Discussion Questions

1. Why are policies regarding the use and allocation of a nation's resources so crucial to securing and maintaining adequate agricultural production?
2. Discuss the support services most needed by the agricultural sector and tell how the governments can best encourage agricultural production through support of these services.
3. How would you describe the assumptions regarding development of U.S. policy toward assisting other countries agriculturally?
4. Differentiate between political and politics.
5. What are the major differences between corporate, group, state, and family farms?
6. List several countries that are experiencing political difficulties that have required military or police enforcement. What impact does enforcement have on agricultural production in those countries?
7. What should be the prime goal of agricultural policy in any nation?
8. Explain why smaller farms, as compared to larger farms, tend to be more efficient operations.
9. Farmers are economically rational. Do you agree? Disagree? Explain your answer.

SECTION 2
Fostering Worldwide Agricultural Development

CHAPTER 10
Institutions and Agencies Providing Assistance and Training Programs

Source: United Nations Development Program; Photo by C. Hart.

Terms to Know
AID
emergency aid
food aid
development aid
economic aid
military aid
FAO
UNDP
Peace Corps
PVO

Objectives
After reading this chapter, you should be able to:
- distinguish between the various types of assistance available.
- identify and explain the nature of assistance provided through international institutions and organizations.
- identify and explain the nature of assistance provided through government institutions and agencies, churches and other religious groups, international voluntary services, and private foundations.

Concepts of Foreign Assistance Programs

In terms of motivation which underlie the efforts of citizens in a nation to assist in overcoming food deficiencies in another country, compassion and genuine concern must be basic. Aid may also be distributed because grantors feel that such action is advantageous to their nation in furthering attainment of such goals as goods, wealth, and power.

The prevailing image of foreign assistance held by grantor nations seems to be one of superior, advanced, skilled people sharing their abundance with inferior, less advanced, less skilled people. However meritorious our sharing may be or may have been, many organizations urge a transition to a higher concept of sharing.

Benevolence and dependency are not a suitable basis for long-term harmonious relationships. Foreign aid as a one-way form of charity is, in the long run, demeaning and resented. Perpetual gratitude for favors bestowed by richer neighbors is not a natural attitude that can be sustained. Is two-way sharing possible between the U.S. and the countries to which it has given assistance? At first glance, that may seem highly unlikely. Yet it is a possibility that must be explored, and is the ultimate goal toward which we must work. Foreign aid must become a partnership, or eventually it will be rejected by both sides.

Information and coordination are the keys to achieving a closer relationship of government and private groups working toward a common agricultural production goal. Each party needs to understand what part of the problem it can address. When an Agency for International Development (**AID**) mission, for example, has a satisfactory plan to develop a country's food system (developed with that nation's involvement), it should be shared with private agencies operating within that country. A properly prepared plan will illuminate the specific needs of home gardens, nutrition training, irrigation, or extension work. Once known, this process may allow a private agency or agribusiness to select a smaller project that it has the resources to handle.

Types of Aid Given

For the task of providing adequate nutrition for all people, programs and projects can be classified into emergency or immediate, and developmental or long-term.

National disasters such as drought, earthquakes, typhoons, or hurricanes, and man-made calamities such as war, all call for **emergency aid** in providing immediate food for survivors. The need for assisted nations to keep themselves adequately fed is of even greater importance when viewed from a global perspective. Programs and projects may also be categorized into direct or indirect.

The terms **food aid** and **development aid** are sometimes used in describing major efforts to assist a nation whose people may be suffering from severe malnutrition, see Figure 10–1. Another distinction may be made through reference to **economic aid** and **military aid**.

Figure 10–1 • U.S. AID airlifts medical supplies to Albania in 1991. The medical supplies were distributed to some 36 urban and rural hospitals, and other health centers throughout Albania. Albania is the poorest country in Europe with a per capita GNP below $700. It is estimated that at least 1500 newborn babies die each year for lack of basic medical equipment, supplies, and pharmaceuticals. *Source: United States Agency for International Development; photo by Clyde McNair.*

Classification of Agencies and Institutions Providing Aid

As mankind becomes more sensitive to the needs of all people, and is confronted with the gross differences in resources between the developed nations and the developing nations, the need to share and coordinate efforts become apparent. Groups providing aid may be classified as follows:

1. organizations and agencies of international scope and management, both bilateral and multilateral.
2. government (national) groups and organizations.
3. churches and religious groups.
4. private and philanthropic foundations and agencies.

The types of assistance given vary among agencies and organizations. Certain agencies, often governmental, function more in the area of policy making and planning, while other groups may be more involved in the actual direct administration of aid. As an example, The World Bank, headquartered at Washington, D.C., functions as the lending agency for a large portion of the developing world.

Food and Agriculture Organization (FAO) of the United Nations

The function of the FAO of the United Nations (U.N.) is to collect, analyze, interpret, and disseminate information relating to nutrition, food, and agriculture. (The FAO is one of 15 U.N. organizations known as a U.N. specialized agency.)

In the FAO constitution, the term "agriculture" and its derivatives include fisheries, marine products, forestry, and primary forestry products (FAO Basic Texts, 1974, p. 4).

The FAO has also served effectively in holding the U.N. Conference on Food and Agriculture meeting every two years. This conference has assumed a very important place in providing a forum for debate and deliberations among member nations. Since the establishment of the FAO, a number of specialized agencies have been developed within the U.N. which serve the cause of agricultural development and food production. Within the U.N., associated organizations have expanded rapidly, based largely on pre-investment projects and technical assistance added to their normal functions of fostering cooperation and research; see Figures 10–2, 10–3, and 10–4.

United Nations Development Program (UNDP)

A major unit of the U.N. global effort accompanying the FAO is the multifaceted UNDP. Supporting agricultural development projects in more than 113 nations, UNDP operates with a governing council of representatives from 50 nations representing all economic and political shades of the spectrum. While UNDP is largely funded through committed FAO funds furnished by U.N. member countries, certain nations may provide supplemental funding for specific projects. In addition, budgets for specific projects often include contributions from a number of agencies. For example, a project directed toward demonstrating the feasibility of soybeans as a useful and marketable crop for selected African nations has the FAO, UNICEF, CARE, and the University of Illinois participating with the UNDP. In one of its publications, the UNDP used the following example to describe its global involvement:

> The priority:
> In the 31 least developed countries alone, some 70 million human beings are clinically undernourished. That figure will double unless vigorous remedies are applied. This is only one of many factors making it imperative

Figure 10–2 • Fertilizer consumption in Bolivia is extremely low. The major reasons are insufficient knowledge of fertilizer use and requirements among farmers, lack of technical assistance, and limited availability of imported fertilizers. In 1986, the Bolivian government requested a project aimed at introducing rational fertilizer use with the objective of increasing land crop production, thus helping to raise the small farmer's income and living standards. Financed by the Government of the Netherlands and executed by the FAO, the operations began in 1987. *Source: Food and Agriculture Organization of the United Nations; photo by A. Conti.*

Figure 10–3 • FAO technical assistance provided to the upper and middle Sao Francisco irrigation project in Brazil, 1990. Pictured is cultivation of sugar beets within this project. *Source: Food and Agriculture Organization of the United Nations; photo by A. Conti.*

that we lose no time in assuring the security and availability of world food stocks on a long term basis.

The projects include research on:

- food systems and policies to identify constraints on food security and the steps required to remove them.
- growing high-yield, protein-enhanced, disease- and pest-resistant strains of rice, maize, sorghum, millet, cassava, potatoes, sweet potatoes, and wheat.
- increasing the genetic diversity of basic crops to enable cultivation in a broad range of soils and climates, and to safeguard against destruction of major species.
- fertilizer production and utilization including development of legumes that can "feed" themselves by employing free living bacteria and algae to fix nitrogen out of paddy soils.

Figure 10–4 • FAO technical assistance provided to the upper and middle Sao Francisco irrigation project in Brazil, 1990. Pictured is a farmer tending his bean crop. Note sprinkler irrigation in background. *Source: Food and Agriculture Organization of the United Nations; photo by A. Conti.*

- preventing and curing livestock diseases such as the animal strain of sleeping sickness that devastates large areas of Africa and South America.
- inventorying the world's renewable marine sources through acoustic surveys; and sampling commercial scale fishing in every ocean on earth.

Officials of the UNDP lay claim to several key attributes and accomplishments, which include the following:

1. the program assists more countries and people than any other program.
2. it is active on more economic and social fronts than other programs.

3. it focuses on priority needs—virtually 80 percent of assistance is concentrated in the poorest countries.
4. its extensive field service network gives the greatest capabilities and working experience in coordinating both in-country and inter-country activities.

Such activities include coordinating and servicing functions for multilateral, bilateral, private sector, and nongovernment organizations.

Other organizations and agencies which have worked with or supported UNDP include the following:

1. United Nations Department of Cooperation for Developing Countries (UND-CDC).
2. United Nations Educational, Scientific and Cultural Organization (UNESCO).
3. United Nations Children's Fund (UNICEF).
4. United Nations Fund for Population Activities (UNFPA).
5. International Fund for Agricultural Development (IFAD).
6. Arab Fund for Economic and Social Development (AFESD).
7. United Nations Volunteers (UNV).
8. World Food Program (WFP).
9. United Nations Conference on Trade and Development (UNCTAD).
10. Development Banks, including the following:
    - World Development Bank (WDB)
    - Inter-American Development Bank (IDB)
    - African Development Bank (ADB)
    - Asian Development Bank (AsDB)

Assistance by National Governments

Each nation has some type of organization and structure for the provision of an adequate supply of food among its people. Some nations have more effective programs than others. While programs administered by the developed nations tend to focus on grants and loans given to developing, often poverty-stricken nations, those efforts by the Third World nations themselves tend to center more on administration and distribution of recipient aid and services. The extent to which donor nations agree to coordinate with and contribute through world organizations varies greatly, but the greatest portion of aid rendered by developed nations is through a U.N. affiliated group. The assistance rendered by the Netherlands, Denmark, Sweden, Belgium, Germany, the United Kingdom, Australia, and Canada can be cited as constituting effective efforts. Recently, Saudi Arabia and Kuwait have given generously to enhance agricultural production and food distribution, particularly to countries with large Muslim populations.

U.S. Assistance Programs

Historically, legislation in the U.S. directed specifically toward assistance to the underdeveloped nations had its beginning in 1950 with the passage of the Act for International Development. This legislation authorized the Point Four Program through establishment of the Technical Cooperation Administration (TCA). In terms of agricultural development, other existing agencies often played leading roles in the planning and implementation phases. These organizations included the Economic Cooperation Administration (ECA) and the Mutual Security Agency (MSA). Sometime later, TCA became known as the International Cooperation Administration (ICA), and still later the official title became Agency for International Development (AID), the name under which it is presently operating. Though agricultural aid was considered to be an important segment of total assistance, often military aid figured more prominently, particularly from the standpoint of a short-term expedient. Let us examine AID more closely.

AID

AID has not only contributed to the welfare of developing countries, but also to the welfare and national security of the U.S. AID has provided development assistance to some 100 countries in total.

It has been said that knowledge of the past prepares us for the challenge of the future. One lesson learned over the years is that development takes time. No matter what the approach, it is complex and lengthy.

AID has traditionally concentrated on capital projects such as dams, railroads, and highways in the belief that they would ignite an economic take-off in the developing countries. This would create jobs, food, education, and higher income for the poorest people in those countries.

While some results of AID have fallen short, the creation of infrastructure has contributed notable achievements. For example, the average Third World life expectancy increased from 35 to 50 years—the level attained in Western Europe at the beginning of the 20th century. Also, the percentage of adults in low-income countries who can read and write has risen from 10 percent to well over 23 percent.

Some developing countries achieved such large economic growth that they no longer need U.S. assistance. Subsequently, they established new economic relationships with the U.S. based on trade and private investment. Taiwan and South Korea are two examples.

In 1988, selected target countries were each allocated $40 million in assistance from AID, and included the following countries:*

1. Bangladesh
2. Pakistan

---

*By law, AID technical assistance programs cannot be provided to countries with a per person gross domestic product (GDP) of more than $1,320 per year.

3. Bolivia
4. Dominican Republic
5. Jamaica
6. Honduras
7. Egypt
8. Portugal
9. Costa Rica
10. El Salvador
11. Africa—regional/selected ad hoc
12. Israel
13. Turkey
14. India
15. Guatemala

USAID Rushes Aid to Kurdish Refugees: An Example of Emergency Food Aid*

After their uprising against Saddam Hussein failed at the end of the Gulf War, hundreds of thousands of Iraqi Kurds fled their lands in fear of reprisal, seeking refuge in the mountains along Iraq's border with Turkey and Iran.

They were stranded without food, water, or shelter. Kurds began to die by the hundreds.

At the President's order, USAID sprang into action as part of the international relief effort.

The Agency's Office of Food for Peace authorized the dispatch of food aid for the refugees. In preparation for this contingency, Food for Peace had prepositioned 29,000 metric tons of food. The total Food for Peace contribution was 49,800 metric tons of food valued at $31.6 million.

A Disaster Assistance Response Team from the Agency's Office of U.S. Foreign Disaster Assistance (OFDA) determined the humanitarian needs of the Kurdish refugees.

The team was composed of experts in the fields of medicine, food, logistics, water, shelter, and communications. Team members coordinated efforts with the U.S. Embassy, the government of Turkey, the Turkish Red Crescent Society, the U.N. High Commission for Refugees, and other voluntary organizations.

U.S. Public Law 480: The Food for Peace Program

Since 1956, legislation known as U.S. Public Law 480 or the Food for Peace program has been in operation. This was a form of direct aid from the U.S. Through provisions of Public Law 480 and its amendments, farm-grown products with a total value of over 33

---

*Adapted from the U.S. Agency for International Development.

billion dollars have been sold, traded, or donated during the past 25 years. In a number of cases, a major portion of payment was made to the U.S. through labor on development projects situated in the recipient country. Among other provisions, the legislation requires that at least 75 percent of food commodities provided be allocated to countries that have a per capita GNP under $300.

In terms of the future of Food for Peace assistance, Bloch (1981, p. 5) made the following comment:

> It is, of course, necessary to be alert to the danger that food aid to developing countries will create a disincentive to agricultural production and a dependency on food imports and foreign aid. Therefore, it is our job to program aid carefully to avoid these dangers. Recipient countries must be convinced that food aid is only a "stopgap," as you say. The food as well as the resources it may generate must be directed toward achieving greater agricultural self-sufficiency. Governments cannot ignore policy and institutional reform in favor of short-term food aid.

Title XII and the Board for International Food and Agricultural Development (BIFAD)

In 1975, when Congress passed the legislation that is known as Title XII, it laid down a challenge for the launching of a new kind of global partnership. The Congress described how various participants working together could make the difference in preventing famine and establishing freedom from hunger for all people.

Title XII is a document based upon experience and hope, and reflects a practical approach. Title XII states the following items:

- that the application of science is a major key to solving the food and nutrition problems of the developing countries.
- that research and its application require long-term support.
- that the capacity to provide technical assistance needs to be strengthened.
- that a dependable source of federal funding is needed if resources are to be used in assisting agricultural production advances in developing countries.
- that new, more effective ways can be developed for AID and U.S. universities to build on and work with existing programs and institutions, such as the international agricultural research centers, the U.S. Department of Agriculture, and other government agencies.
- that work directly related to agricultural production in the developing countries should largely be carried out in those countries and adapted to local needs.
- that the developing countries need their own institutions and trained people to carry out research, extension, and teaching activities.
- that our food and agricultural efforts in developing countries can benefit the U.S. as well.

The Congress therefore directed the President to establish a permanent Board for International Food and Agricultural Development (BIFAD). This Board was charged to work with the Administrator of AID in carrying out the programs authorized by Title XII (Amendment to the Foreign Assistance Act of 1961) because BIFAD is part of IDCA, not part of AID. In 1990, BIFAD's name was changed to Board for International Food and Agricultural Development and Economic Cooperation (BIFADEC).

Title XII in the Developing Countries

Title XII programs in the developing countries focus on agriculture, aquaculture, nutrition, agriforestry, and closely related fields. Primary objectives include the following:

- the development of the country's capacity to do research, extension, and teaching.
- the training of people at all levels to carry out and continue agricultural development work adapted to local needs and circumstances.
- the discovery of new knowledge through the conduct of research.
- the improvement of local systems to deliver the best available knowledge to small farmers and farm laborers, many of whom are women.

Title XII in the U.S.

Title XII activities in this country focus on the capacity of U.S. agricultural universities to supply AID with the support needed in its overseas programs. Primary objectives include the following:

- the strengthening of U.S. university curricula in such areas as language training, and the specific fields of expertise needed by AID in its programs overseas.
- the encouragement of participation by top scientific and other professional talent in AID's programs.
- the development of interest by U.S. students in advanced training and work in the areas of international agricultural research, extension, and teaching.
- the development of effective procedures for locating and acquiring the needed skills and resources identified by AID.
- the establishment of research programs that focus on the problems of the developing countries and that emphasize scientist-to-scientist relationships.

Volunteer Agencies

The Peace Corps

In many nations, individuals serve as volunteers in assisting the people of developing nations to achieve self-sufficiency. In the U.S. such a government-sponsored program for sending volunteers overseas to serve in community development programs is the Peace Corps.

Volunteers were sent overseas under a set of institutional goals provided by Congress in the Peace Corps Act (Public Law 87-293) of September 22, 1961. Three broad objectives were stipulated by Congress as follows:

- to help the countries inviting volunteers to meet their needs for trained manpower.
- to promote abroad a better understanding of Americans and American society.
- to promote in the American people a broader understanding of other people.

The Peace Corps still serves as an opportunity for idealistic youth to invest their commitment, although a number of older volunteers and retired persons have contributed substantially also.

While not specifically named as such, the Peace Corps can be rightfully recognized as an agency of change. The idea of doing something positive and concrete to effect change and to solve problems in the Third World is an implicit assumption of almost everything the Peace Corps has done.

The International Executive Service Corps (IESC)

Another unique voluntary group from the U.S. is the International Executive Service Corps (IESC), an AID-sponsored group composed of business persons and managers who volunteer services to their counterparts in developing countries. More than one third of all assignments have been for enterprises in agriculture and food processing, construction and building materials, and apparel manufacturing. Recently, there has been more emphasis on health care and education.

Assistance Programs by U.S. Universities

While contract aid provided by U.S. universities will be analyzed in detail later in the text, it should be noted that many member institutions of the National Association of State Universities and Land Grant Colleges have provided assistance to developing nations in the area of agriculture, particularly through sponsorship by AID and its predecessors. Title XII programs provide for U.S. universities to be directly involved with their counterparts in developing countries.

It is the policy of U.S. government to mobilize university resources to carry out technical assistance projects. Nearly all university activities are U.S. AID funded; therefore, little of the university's funds are used for these purposes.

State Universities and Land Grant Colleges have taken up the challenge of helping bring productive agriculture wherever the need may be. Although colleges and universities have been active participants in providing technical assistance for years, recently they have been requested to provide assistance to emerging nations such as Bulgaria and Poland. Another example is Mexico's interest in obtaining assistance in improving agricultural education, particularly within the university system in the State of Veracruz. Initial concepts include the following:

1. provision for short courses and workshops conducted by U.S. faculty.
2. improving the existing agricultural program at the University of Veracruz through faculty development and curriculum development.
3. establishment of a new university that will focus on production agriculture under tropical conditions.

Other Organizations and Agencies

Churches and Religious Organizations

The total effort directed toward emergency and developmental hunger alleviation programs by churches and religious organizations is substantial. Examples of denominational and inter-denominational Christian church groups are the Church World Service (CWS), Lutheran World Relief (LWR), and World Relief Commission (WRC). The Catholic Relief Services (CRS) serves in an effective manner in behalf of those of the Roman Catholic faith. Many denominations have units within their church which are specifically charged with responsibility for world hunger as a church priority. The United Methodist Church has an active subgroup within its Board of Global Ministries known as UMCOR (United Methodist Committee for Overseas Relief). Another example of an effective program is one administered by the Christian Reform Church. Through their Christian Reformed World Relief Committee, carefully monitored programs in 20 selected nations are centered on the following:

- food production.
- health and sanitation.
- income generation.
- literacy.
- community organization.
- rehabilitation.
- refugee assistance.

A number of groups claiming to be non-denominational are actually supported by single churches as well as ecumenical units. Examples of these include the following:

- International Voluntary Services.
- Oxfam.
- World Neighbors.
- World Vision.
- Heifer Project International.
- Christian Rural Overseas Program.
- Bread for the World.

Private Organizations and Agencies

In the U.S., Canada, and in some European nations, a number of agencies serve effectively in the struggle against world hunger. CARE, for example, is the world's largest private, nonsectarian dispenser of foreign aid. It has continuing efforts in 36 countries, with programs of great diversity. While feeding the hungry is still a major focus, the spectrum of related side activities may range from sanitary latrine construction to the distribution of dried milk and fortified cereal preparations; see Figure 10–5. Another unique organization is Africare, with efforts largely directed toward providing relief and development assistance to the poorest of African nations.

Currently, over 21,000 foundations are in existence in the U.S., with many others in Canada and Western Europe. However, no more than 150 U.S.-based foundations can be identified as making grants that directly support international program development. Many others, however, do grant funds to **PVOs** (Private Voluntary Organizations) that spend the bulk of their funds outside the U.S.

It is alleged that the PVOs, by their very nature, possess certain advantages over government or quasi-government agencies for carrying out the tasks of hunger alleviation and development. PVOs are particularly active in fostering self-help initiatives among the poor. PVOs are capable of mobilizing substantial financial and human resources, not only among U.S. citizens but also among the citizens of the host country. Finally, PVOs strengthen people-to-people contact, whereas government programs sometimes replace personal relationships with impersonal institutions, thus undermining ties between people.

Various PVOs have been forced to recognize that frequently their fund-raising efforts are viewed by potential donors as random in nature, sometimes competitive, and too often poorly coordinated. They must be successful in maintaining an image as nongovernmental entities and, therefore, more acceptable than official U.S. agencies. PVOs also desire increased independence from federal policies, but at the same time insist that there should be a sizable increase in funds for co-financing with AID or other government agencies.

Figure 10–5 • A woman tends to beehives in rural Belize. CARE helped her launch a career as a beekeeper and escape poverty with training and a loan to purchase equipment. *Source: CARE; photo by Steven Maines.*

Conclusion

It is believed that, as nations develop, they will become more stable and democratic as the ills of poverty, food shortages, and unemployment are reduced through development projects. The various institutions and agencies discussed in this chapter have as their primary objective the improvement of political and social stability via enhanced agricultural and economic assistance.

Assistance programs have engendered progress in a number of areas. Schools, hospitals, medical centers, water and sewage systems, and housing complexes have been erected. Factories and industries have been built. Roads, airfields, port facilities, dams, and irrigation systems have been constructed. Advanced technologies have been introduced, illnesses eradicated, and food production increased.

Without assistance from donor nations, private foundations, private voluntary organizations, and other agencies, the world would suffer from more poverty, hunger, illiteracy, and illness. Yet despite the levels of aid and assistance already given, these conditions still persist; therefore, continued political violence and instability between developing and emerging nations seems likely.

Discussion Questions

1. What are some of the major project areas which UNDP seeks to emphasize in its programs?
2. What other international agencies are often associated with the UNDP in carrying out objectives?
3. Explain the difference between immediate (emergency) aid and developmental (long-term) aid.
4. Briefly describe the purpose of the U.S. Peace Corps.
5. List some development agencies which are
   a. international
   b. national
   c. religious
   d. private
6. What is the primary function of the Food and Agriculture Organization of the United Nations?
7. List the countries which are the primary recipients of U.S. foreign assistance programs.
8. One of the objectives of Title XII was to develop interest among U.S. students in advanced training and work in the areas of international agricultural research, extension, and teaching. What is your occupational interest and why do you believe you need to be knowledgeable concerning international activities?

CHAPTER 11
Financing Programs for Agricultural Production and Development

Source: DeKalb Plant Genetics.

Objectives

After reading this chapter, you should be able to:

- list the three major requirements for an effective rural financial system.
- identify and discuss the various kinds of production and development projects for which financing is needed.
- identify and describe the roles of lending agencies in furthering rural development and agricultural production.
- discuss the operations and priorities of development banks.

Terms to Know

World Bank
International Bank for Reconstruction and Development (IBRD)
International Development Association (IDA)
Inter-American Development Bank (IDB)
International Monetary Fund (IMF)
Asian Development Bank
European Development Fund
Export-Import Bank of U.S.
development banks
commercial banks

Rural Finance—Requirements for an Effective System

Credit becomes important as a rural community moves from traditional to modern stages of development—from a barter to a cash economy. Inputs must often be purchased with cash. A surprising amount of capital can be mobilized by the farmers themselves, either from their own funds or from informal sources—relatives or moneylenders—in response to profitable opportunities. However, many farmers, especially those with smaller operations, may not have access to adequate financial resources and must use other sources.

Credit enables the production process for those unable to provide their own financing. Credit can also help people adjust to risk and uncertainty. The compelling incentive to borrow should be the expectation of higher profits—more than enough to cover the costs of the credit.

There are three basic requirements for an effective rural financial system. First, farmers must have profitable investment opportunities. Such opportunities will exist if improved technologies, reasonably priced production inputs, attractive product prices, and accessible and stable markets are available.

Second, interest rates should be at or near levels of those in the commercial market. Interest is a cost—the cost of capital. The higher the interest rate, the higher the economic returns to cover this cost must be. The interest rate, like the cost of any other input, serves as an allocating mechanism, in this case encouraging or discouraging investments which have a capital component.

Third, financial services should be made available to all persons. Rules governing rural credit should be no different from those governing credit in general. The main differences are ones of degree—the nature of the risks, the seasonality of production, and the numbers and sizes of the borrowers.

Farmers Require Financial Assistance

Though the vast majority of farmers in other countries operate only small farms with very limited and sometimes primitive equipment, their farming operations nevertheless require some degree of financing. As farmers around the world adopt improved practices and make use of appropriate technologies, their need to secure adequate financing becomes correspondingly greater. Larger programs of development that involve agricultural areas call for big expenditures that are often under the control of governmental units. Joint efforts with the United Nations Development Program or with bilateral operational agencies such as the U.S. Agency for International Development are examples of such large-scale programs. Commercial banks, particularly those in the U.S.,

Figure 11-1 • There is a rising interest in the production and consumption of fresh vegetables in Cape Verde (1989), which is encouraged and supported by the government. Pictured is an open-air conservation of seed potatoes on trays in small wooden huts. The project was funded through the Food and Agriculture Organization's Technical Cooperation Program with additional assistance from the Government of Austria. *Source: Food and Agriculture Organization of the United Nations; photo by J. M. Micaud.*

extend medium-term credit to governments and other development agencies. Lesser amounts are made through many PVOs. Financing of both production and development operations can be viewed as a colossal task. From the distribution of loans and subsidies to individual farmers, to the allocation of funds to countries through multilateral or bilateral agencies, the process is often complicated by political considerations, as was pointed out in Chapter 9; see Figures 11–1, 11–2, and 11–3.

Figure 11-2 • Women weeding in the vegetable-growing area of Saint Domingo on the island of Santiago. This vegetable-growing effort serves as a trial, for reference, and to teach the local populace. The project was funded through the Food and Agriculture Organization's Technical Cooperation Program with additional assistance from the Government of Austria. *Source: Food and Agriculture Organization of the United Nations; photo by J. M. Micaud.*

International Programs of Financial Assistance

United Nations Development Program (UNDP)

A major dispensing agency which provides assistance for development programs is the UNDP. As a related part of the United Nations World Food Organization (WFO), UNDP lists the following areas, each brought about through mandates and resolutions of the United Nations:

- combatting hunger
- tax reform

Figure 11–3 • Joygun (right), an impoverished widow with three children, sells rice in her small village in Bangladesh. A loan from CARE helped her launch a small business as a street vendor. *Source: CARE; photo by George Wirt.*

- tourism
- protein nutrition improvement
- public administration
- drought recovery and rehabilitation
- promote economic cooperation
- hydrological surveys
- conservation methods
- solar energy sources
- human settlements development
- primary and secondary education
- natural resources development
- scientific and technical development

- mineral resources exploration
- implementing global projects
- raw materials marketing
- combatting desertification
- developing island countries
- integrated rural development
- irrigation surveys
- agrarian reform
- technology transfer
- geothermal sources
- water resources management
- disaster relief
- improving educational systems

The role of national governments in granting and receiving funds can be summarized as follows:

- furnish financial resources through voluntary contributions (industrialized and developing nations).
- establish policy guidelines for country and intercountry resource allocations through rotating service on the Governing Council and at the United Nations General Assembly (industrialized and developing nations).
- set priorities for UNDP assistance in their own countries and provide over one-half of local project costs.

The World Bank

There are several banks that make loans throughout the world. One of the largest is the **World Bank,** which consists of the **International Bank for Reconstruction and Development (IBRD)** and the **International Development Association (IDA).** The World Bank has more than 127 member nations. The Bank was founded in 1944 and opened for business in June 1946. On the average, bank loans are repaid over a period of about 20 years at a relatively low interest rate. Each of the Bank's loans is guaranteed by the government concerned.

The volume of annual disbursements on project lending is influenced by the following factors:

1. the complexity of projects.
2. economic conditions in the borrower countries.
3. availability of expertise.

4. the size of the undisbursed balances on approved loans and credits.
5. the sector and age of the undisbursed balances.

Examples of annual project lending disbursements are presented as follows:

- Bangladesh borrowed monies in order to assist rural cooperatives in offering services and facilities for tubewells, low-lift pumps, and warehouses.
- Kenya borrowed monies in order to introduce a national agricultural extension service. Approximately 1.7 million farm families were to be the beneficiaries.
- Jamaica borrowed monies in order to finance imports of raw materials, intermediate goods, nonluxury consumer goods, capital equipment, and spare parts. These imports were meant primarily to stimulate industrial and agricultural production for export.
- Kosovo (a sector of Yugoslavia) received monies to provide farm credit. It was a loan directed primarily at promoting Yugoslavia's individual agricultural sector development.

Inter-American Development Bank

The **Inter-American Development Bank (IDB)** was established in 1959. It is a major instrument in channeling loan assistance to 20 Latin American members. In 1964, the capital stock was increased. Recently, it has been reported that the bank had made in excess of 167 loans for over $1.7 billion into the region's agriculture. It has made other loans, but of particular interest and concern is the amount of monies placed into the agricultural sector.

International Monetary Fund

The **International Monetary Fund (IMF)** began operations in Washington, D.C. in 1946. Like The World Bank, the IMF has over 100 member nations. Membership in the IMF is a prerequisite for membership in The World Bank. The IMF is financed by subscriptions from its member nations. The IMF has gradually developed into what is now the world's most powerful international economic institution. It has become the institution which leverages lending by both official aid agencies and private commercial banks—neither of whom will lend for long to governments out of favor with the IMF. Thus, the IMF has vast power not only to determine which countries qualify for external loans, but also to dictate national economic policies to borrowing governments.

Asian Development Bank

The **Asian Development Bank** started operations in 1966. The bank has 32 member nations. Most of the bank's loans are hard loans. It should be noted that the bank is seeking additional resources to enable it to make soft loans.

European Development Fund

The **European Development Fund** was created in 1958 by the six members of the Common Market. The primary purpose of this Fund was to promote economic and social development of the overseas countries and territories associated with the member countries.

Nature of U.S. Foreign Assistance*

U.S. foreign assistance has been evolving since World War II. U.S. assistance began as assistance that could be called economic, technical, or developmental. Its goal was to improve the economic welfare of those in other countries.

The rise of the Cold War added a new twist, driven by political objectives in the name of security assistance. Since then, the U.S. has included nations deemed important to its foreign policy strategy. Iran, Vietnam, Israel, and Egypt have all been major recipients of economic and security assistance.

U.S. foreign assistance today is composed of economic and military aid; the former is a mix of developmental objectives in the poorer countries, and security interests worldwide.

Origin of U.S. Foreign Assistance

Prior to 1949, the U.S. had only dabbled in international development. In 1942, the Institute of Inter-American Affairs was established to implement the first technical cooperation program supported in modern times by the U.S. In 1948, the U.S. embarked on providing large amounts of capital for war reconstruction efforts with its Marshall Plan in Europe. The overwhelming success of this plan enticed the U.S. to contemplate a broader effort among other countries, and to create the "Point 4 Program" that later evolved to become the U.S. Agency for International Development (AID).

The U.S. has since become the major donor in total assistance offered (but not in terms of percent of GNP in comparison with several other countries). As a world power, the U.S. has developed a tradition of assistance and leadership. Every administration since 1949 has supported foreign assistance as an essential part of this nation's commitment abroad.

As the world has changed and continues to change, countries have become more interdependent and the need for international cooperation has increased.

---

*Adapted from the National Association of State Universities and Land Grant Colleges.

International Agricultural Development Lending Institutions in the U.S.

The Export-Import Bank of the U.S.

The **Export-Import Bank** (Eximbank) is the U.S. Government's oldest (1934) agency to guarantee foreign lending. During the 1960s, approximately $600 million per year was loaned for development purposes. Latin America is the largest recipient of long-term Eximbank loans. Most of the bank loans must be spent for commodities or services in the U.S., or to pay off private credits in the U.S. arising from previous purchases.

The bank is a wholly U.S. Government corporation and sells participation certificates in loans, guarantees, and insurance.

The bank underwrites 100 percent of the political risks and an agreed-on percentage of an export credit insurance plan of the Foreign Credit Insurance Association. This insurance covers the exporter, up to an agreed percentage, against nonpayment loss resulting from commercial credit, political hazards, or both.

Agency for International Development

The Agency for International Development (AID) is an agency of the Department of State and administers the U.S. foreign economic program. AID came into existence in 1961 when it took over the functions of the International Cooperation Administration and Development Fund. Some of the most important activities of AID are investment guarantees, dollar loans, local currency loans, and investment surveys.

The purpose of the guarantee program is to encourage U.S. private investors to invest in the less developed nations. U.S. dollar loans are usually made to cover the part of the project procured by the U.S.

AID receives its funds from annual appropriations of the U.S. Congress. Up to 25 percent of the foreign currencies received from sales of agricultural commodities under Public Law 480 may be loaned to U.S. firms for business development and trade expansion in a foreign nation.

Most AID programming has stressed agriculture and rural development. AID development assistance falls into four primary categories:

1. agriculture and food.
2. population, nutrition, and health.
3. rural development.
4. public administration and policy.

Each of the aforementioned categories involves a wide range of subcategories, which generates even greater diversity of individual projects. Some of the priority concerns of AID include the following:

- functional development
- Sahel Development Program
- American schools and hospitals abroad
- disaster relief
- operations and maintenance
- Foreign Service Reserve
- trade and development programs
- International Narcotics Control
- Inter-American Foundation
- Peace Corps
- Africa Development Foundation
- migration and refugee assistance
- Public Law 480 (Food Aid)
- Miscellaneous Trust Fund

In the earlier years of AID, focus was given to capital-intensive projects such as irrigation development, road construction, communications, and rural electrification. More recently, the emphasis has shifted to projects with social dimensions and human development. Institutional building, agricultural research and extension, family planning, nutrition and health, policy dialogue, more involvement of the private sector, and market forces are the issues of development being stressed currently.

Since 1946, some 150 nations have received U.S. foreign assistance (Table 11–1). About 75 nations currently receive U.S. aid., with most in Asia and Africa. Fewer nations receive assistance in the Middle East, while support for Latin American countries has been reduced considerably in recent years. Nations in Central America and the Caribbean are receiving more attention, but the amounts of money compared with other regions are small.

Among individual nations, Israel and Vietnam top the list of all-time recipients (this aid was mostly for security assistance). South Korea, Egypt, and Turkey have also received large amounts, also for security assistance. The major recipients of developmental assistance have been India, Pakistan, Indonesia, and the Philippines. Today, Bangladesh receives the largest amount of nonsecurity assistance of any nation.

U.S. Development Banks

Since 1945, national **development banks** and corporations have played an increasingly important role in stimulating the growth of economies in the less developed nations. These development banks supply needed capital to furnish long-term financing to

Table 11–1 •

Top 30 Nations That Have Received U.S. Foreign Assistance

1.	Israel	16.	Morocco
2.	Egypt	17.	Honduras
3.	Turkey	18.	Jordan
4.	India	19.	Somalia
5.	Greece	20.	Jamaica
6.	Pakistan	21.	Tunisia
7.	Spain	22.	Kenya
8.	El Salvador	23.	Costa Rica
9.	Sudan	24.	Peru
10.	Republic of Korea	25.	Sri Lanka
11.	Indonesia	26.	Dominican Republic
12.	Bangladesh	27.	Liberia
13.	Philippines	28.	Lebanon
14.	Portugal	29.	Zimbabwe
15.	Thailand	30.	Oman

supplement local capital markets. These banks also play an important role in providing technical assistance to the local investors. Most of these development banks are government controlled and subsidized.

The World Bank has made substantial loans to the development banks and has provided much-needed technical advice.

U.S. Commercial Banks

U.S. **commercial banks** are actively cooperating with various U.S. governmental and international lending agencies. Some of the commercial banks extend medium-term credit to foreign governments, central banks, commercial banks, or selected private corporations to finance development projects. These banks maintain close working relationships with the U.S. government and foreign lending agencies, and are able to provide expert advice to their customers on many problems relating to international trade.

Conclusion*

The U.S. has a large capacity for assisting and a high level of willingness to do so.

---

*Adapted from the National Association of State Universities and Land Grant Colleges.

Economic assistance programs generate benefits for the U.S. as well as for farmers around the world. There are compelling economic and security reasons for the U.S. to engage in such assistance.

Agricultural growth can contribute to overall global economic growth in the following ways:

1. agricultural profits deposited in financial institutions can in turn lend funds to nonagricultural and agricultural sectors.
2. higher agricultural incomes provide a source for taxes.
3. raising a country's foreign-exchange earnings through increased agricultural exports, reducing imports, or attracting foreign loans, grants, or investments.

Technological advances alone cannot secure the sustained agricultural change that will benefit large numbers of farmers throughout the world. Institutional framework and economic incentives must encourage the adoption of improved technologies to the point of achieving the full potential gain. Agricultural financial service organizations must be responsive to the needs of producers locally, nationally, and internationally.

Discussion Questions

1. Name the various sources of funding available for agricultural production and development.
2. Give some examples of UNDP development aid and priorities.
3. Summarize the role of national governments in granting and receiving funds.
4. What are some of the influences which affect the lending practices of the World Bank?
5. Why do farmers, around the world (U.S. included) require financial support services?
6. Which two nations top the list of all-time recipients of U.S. financial assistance? Which countries have been the all-time recipients of developmental assistance?
7. List three requirements for an effective rural financial system.
8. Why does credit become so important as a rural community moves from traditional to modern stages of development?

CHAPTER 12

Human Resource Development—A Key to Development of Modern Agriculture

Source: CARE; Photo by Rudolf von Bernuth.

Terms to Know
Maslow's Theory
survival needs
safety needs
social needs
esteem needs
self-actualization
change-agent
principles of teaching
adoption process
Humanistic-Democratic-
    Participative (HDP) systems

Objectives
After reading this chapter, you should be able to:
- discuss the importance of human resource development with regard to modern agricultural practices.
- thoroughly discuss learning both as a process and as an activity.
- discuss the Twelve Principles of Learning.
- explain Maslow's Theory of Hierarchy of Human Needs.
- discuss the Principles of Teaching that are appropriate for change-agents.

- relate how the traditional adoption process concept should be applied to foster adoption of new ideas and practices.
- describe the role and function of the change-agent in technology-transfer and adoption of new ideas and practices.

> "Any definition of a successful life must include serving others."
> President George Bush
> March 1, 1991

Human Resource Development*

Agricultural production is influenced by the quantity and quality of human resources. The importance of labor availability in agricultural production has long been recognized. People make land and other resources productive. Many countries have yet to provide adequate training facilities and opportunities to create a skilled agricultural work force. In many of these countries, illiteracy is high, skill training low, and public schools inadequate. The capacity of people to provide quality labor and to command good employment options is severely limited. Studies suggest that the acceptance of new agricultural technology is greater with farmers who have had more education.

There is a pressing need to address illiteracy and provide job skills for large portions of a population. Without these abilities, people's options are narrowed, and their capacity to earn income is restricted. Agricultural production and development is often curtailed by an unskilled labor supply. Strong evidence suggests that improved education results in development or modern agriculture. Educated people also tend to adopt family planning methods. Investments in basic and technical skills will greatly improve the potential for increasing all goods and services in these economies.

The World Conference on Agricultural Education and Training held in Copenhagen (1970) as reported by Okatahi and Welton (1985), established that the teacher is the most important aspect of education. Without good teachers, agricultural education as a whole cannot function effectively.

Learning and Motivation

Any program that seeks to solve the problems associated with agricultural production should consider the concepts of learning and motivation.

Technology-transfer and the adoption of appropriate practices and management are dependent on the producer's acquiring the needed knowledge and skills, and in developing a favorable attitude and a personal desire to make the changes.

---

*Portions adapted from the National Association of State Universities and Lang Grant Colleges.

Four relevant points should be considered to develop these favorable attitudes. These are:

1. the user's world is the only sensible place from which to consider adoption of an innovation.
2. the utilization process includes a diagnostic phase where the user's needs are considered and translated into a problem statement (or transfer strategy).
3. the role of the outsider (change-agent or extension agent or teacher) is to serve as a catalyst in the diffusion process.
4. self-initiation by the user creates the best climate for lasting change.

Education: Teaching and Learning

One definition of education is "the process of training and developing knowledge, skill, mind, and character." Education may also be defined as the production of changes in the behavior of an individual. Effective education includes the following:

- contributes to an individual's knowledge.
- helps to improve skills.
- develops desirable attitudes.

Learning is the process by which individuals, through their own activity, become changed in their behavior, their ways of thinking or doing, or their ways of feeling. Three broad categories of change include the following:

1. changes in the things known, or knowledge.
2. changes in the way things are done, or skills.
3. changes in feeling, or attitudes.

Stated another way, learning is any activity engaged in that causes that person to be different afterward.

There is general agreement that learning is an internal process that engages a person's whole being—including intellectual, emotional, and physiological functions. Learning can be described psychologically as a process of need-meeting and goal-striving.

Individuals are motivated to learn to the extent that they feel a need to learn, or to perceive a personal goal that learning will help to achieve. They will invest their energy in making use of available resources (including teachers and readings) to the extent that they perceive them as relevant to those needs and goals.

Learning is an active process. We learn by doing the thing to be learned. Consider what percentage of new practices individuals learn in comparison with how they learn them, as follows:

- 10 percent of what they read is retained.

- 20 percent of what they hear is retained.
- 50 percent of what they see and hear is retained.
- 70 percent of what they say as they talk is retained.
- 90 percent of what they say as they do something is retained.

Learning stems from an interest in or a felt need for something. Individuals learn because they perceive that the possible benefits are greater than the cost of changing or continuing in the same way. There are many motives for changing one's behavior. Some of these motives would include profit, prestige, beauty, fear, health, admiration, social acceptance, personal safety, security, physical comfort, and self-fulfillment.

Let us examine the Twelve Principles of Learning as follows:

1. We learn more readily when we are ready to learn and have a strong purpose or desire to learn.
2. Learning requires motivation, since interest is necessary for effective learning.
3. Learning is more likely to take place if students have a reasonable chance of achieving early success in their endeavors.
4. Learning is simplified if what we are learning is built on something we already know.
5. To be effective, learning must proceed in a logical order.
6. More effective learning takes place when learning impressions come through multiple senses.
7. Good learning results when there is some means of applying what we learn.
8. The most effective learning results when there is an immediate application of what is taught.
9. The more often we use what we have learned, the longer it is retained.
10. Learning is problem solving, and problems must be challenging to stimulate learning.
11. Feelings and emotions are strong incentives for learning.
12. The first learning impressions are usually the most lasting; therefore, it is important not to convey wrong impressions that must be corrected later.

Maslow's Theory of Hierarchial Needs

A number of years ago, a prominent social-psychologist, A. H. Maslow, developed a theoretical model that he presented as a "Hierarchy of Human Needs." Subsequent research in the fields of educational psychology and sociology have confirmed the validity of this model and affirmed its practicality. **Maslow's Theory** is a time-tested and useful model, and is presented in Figure 12–1.

In considering those learners who are citizens of developing nations, the vast majority of them are, of necessity, concerned with the first two levels of Maslow's hierarchy: physiological or **survival needs** and **safety needs.** However, even in

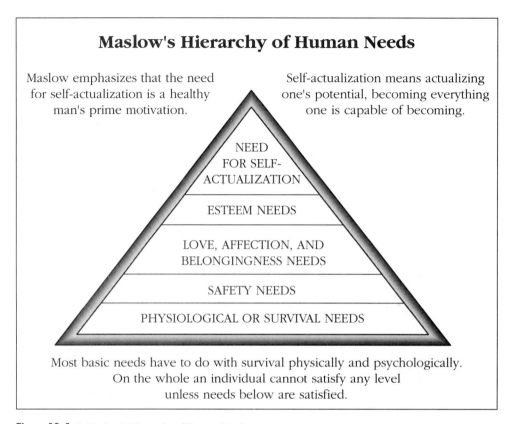

Figure 12–1 • Maslow's Hierarchy of Human Needs.

situations where needs at the lower levels are not adequately met, individuals may profit from awareness of their personal needs in areas placed higher in Maslow's model. Efforts in assistance to developing countries should first be directed toward helping achieve needs identified at the base of the model.

According to Donaldson and Scannell (1986, pp. 68–69)*, each of the levels of Maslow's theory can be explained in the following manner:

1. Level one—physiological needs. Foremost are the biological and physiological things that are needed to survive. These include the basic drives such as food, rest, drink, and shelter.
2. Level two—safety and security needs. When first-level needs are satisfied, second-level needs come into play, and include freedom from fear, danger, and threats. If certain policies that arouse fear or uncertainty are encountered, these needs may become powerful motivators.

---

*Les Donaldson/Edward Scannell, Human Resource Development, The New Trainers Guide, (c) 1986, By Addison-Wesley. Reprinted with permission of the publisher.

3. Level three—**social needs.** Third-level needs deal with social issues. The need for belonging and peer acceptance is important. We want to give and to receive friendship.
4. Level four—**esteem needs.** Personal recognition, such as personal wealth and self-esteem, is the fourth item in the hierarchy. A pat on the back for a job well done, or a word of praise given in the presence of others are important for fulfilling this need. These are also called ego or status needs.
5. Level five—self-actualization. The capstone in human needs is **self-actualization.** The need of self-fulfillment is one that few ever completely satisfy, but it can be a constant motivator since most people will never attain it. However, the need to keep trying tempts one to keep moving toward this goal.

The Change-Agent As Teacher

It has become convenient to use the term **change-agent** to name those individuals whose major role is interpreting, counseling, demonstrating, encouraging, and assisting their students to conceptualize, understand, acquire skills, plan, and put into operation systems that enhance their lives.

Lionberg and Guinn (1982, p. 1) explained that the change-agent's purpose is to help farmers apply new technology. By utilizing newly discovered methods, the farmers can grow more food and support their families better.

Change-agents may also be known as extension agents. Programs of development sponsored and supported by private volunteer organizations and churches may refer to such people as promoters, field workers, or community developers. The majority of such positions are filled by nationals assisted and directed by a few experts from other countries. In the past, when larger programs of aid and development were considered, such as those of the United Nations Development Program (UNDP) and the U.S. Agency for International Development (AID), both top-level and middle-level change-agent personnel were likely to be from other than the recipient country.

The following **Principles of Teaching** are appropriate for all change-agents and promoters:

1. The change-agent accepts clients as persons of worth, and respects their feelings and ideas.
2. The change-agent helps clients diagnose the gap between their aspiration and their present level of performance.
3. The change-agent helps clients identify the life problems they experience because of the differences in their personal farming equipment and other resources.
4. The change-agent involves clients in a mutual process of formulating learning objectives in which the needs of the clients, institution, change-agent, subject matter, and society are all taken into account.

5. The change-agent contributes his/her own feelings and resources in a co-learner role in the spirit of mutual inquiry.
6. The change-agent gears the presentation of his/her resources to the levels of experience of his/her clients.
7. The change-agent shares thoughts about options available in the design of learning experiences and the selection of materials and methods, and involves the clients in deciding among the options.
8. The change-agent helps the clients to apply new learning to their experience, thus making learning more meaningful and integrated.
9. The change-agent involves clients in developing mutually acceptable criteria and methods for measuring progress toward the learning objectives.
10. The change-agent helps clients develop and apply procedures for self-evaluation according to these criteria.

The central responsibility of change-agents is to create and maintain a learning atmosphere that will maximize the acquisition of those concepts, knowledge, skills, and practices by clients. The change-agent can be considered somewhat of a manager of the learning process, but above all, he or she must serve as a functional interpreter; see Figures 12–2 and 12–3.

The list of characteristics that are essential for success as a change-agent includes the following:

- freedom from the handicap of self-centeredness.
- genuine confidence in personal ability as a member of a helping profession.
- foremost, a commitment to a concept that people can be helped to change for the better.
- possession of the concept that, as a change-agent, he or she also continues as a learner.
- development and maintenance of a strong belief that the individual is of supreme worth.

The Adoption Process

Sociologists and professional educators agree on the concept that when individuals are provided the necessary information about any innovative procedure, and when they make an effort to consider change, they quite often follow certain recognized steps that will eventually lead to the adoption of the procedure.

▶ **NOTE** The following description of the **Adoption Process** is credited to Dr. Robert F. Reisbeck, Extension Communication Training Specialist and Associate Professor of Agricul-

Figure 12–2 • Patterson Semenye provides training for farmers in providing improved forages as part of the dual-purpose goat project in Kenya. Semenye is a specialist with Winrock International. *Source: Winrock International; photo by H. A. Fitzhugh.*

tural Education at Oklahoma State University. The material is excerpted from Communications Training Bulletin, No. 7. (More recent and complex models concerning the innovation-decision process have been developed by rural sociologists; however, for purposes of this text, the following model was deemed most appropriate.)

Stages by Which the Learner Makes Changes That May Lead to Adoption

An individual learns and changes in easily-defined stages. Because we can isolate and define these stages, the educator can take steps to help the learner go through each one. These stages include awareness, interest, evaluation, trial, and adoption.

Stage One—Awareness. Simply becoming aware of an idea, a role to be played, an innovation, a way of doing something, or a skill.

When a person becomes aware of something new, recall can occur.

Figure 12-3 • Social forestry in Indonesia. The community organizer (with hat against tree in center) is giving agriforestry training. He received technical guidance from a Winrock International forestry specialist. *Source: Winrock International; photo by Sarah Warren.*

It's easy to become aware of something. It can be done by reading, hearing, or seeing it.

But simply becoming aware of something doesn't necessarily mean that the individual is going to take any action. For example, we are all aware of the need for financial support of church missionaries, yet not everyone will contribute to them.

A person is most likely to become aware of something if there are enough messages to ensure his or her awareness of the new idea.

For the individual learner to retain the message, it is necessary for the learner to realize how the new idea fits his or her situation or problem. So in our educational messages designed to create awareness, we need to show clearly how the information fits into the learner's situation and how it can be of help to the learner.

If the information seems to be useful, the learner may go on to the next step.

Stage 2—Interest. Interest is an outgrowth of awareness. But in the interest stage, the individual begins to search actively for further information about the new idea, role, innovation, or skill.

For the learner's interest to be encouraged, more complete information is necessary. The educator must be conscious of this fact, and continue the messages that were started in the awareness stage. Information must be easily available, and the educator must stand by ready to provide personal help.

If the idea fits the situation, and the information seems rewarding to the learner, that learner may go on to the evaluation stage.

Stage 3—Evaluation. By now the learner may begin to evaluate the information that interests him.

In evaluating the information, some of the questions that the learner may need answers to include the following: "Will it work?"; "How well will it work?"; "Can I do it?"; "How difficult is this?"; "What if I don't do it?"; "Are others doing this?"; "Can I afford it?"; "What other things will I have to do in order to do this?"; and, "Will it take too much of my time?"

The social situation in which the learner finds himself may be another factor. He may ask, "If I do this, what about my neighbors and friends? What will they think? What about those people who are influential in this area? Will they think it's the right thing?"

Without positive answers to these questions, the learner may very well drop the whole matter. Technical or social failure may pose such a threat that it would be best to simply forget it.

It is evident that the educator must be aware of what the learner is going through in this stage. The educator must begin a more personal relationship with the learner in providing answers to questions. Likewise, the educator needs to be aware of the community social situation, and be sure that community influentials and peers can legitimize or provide assurance that the change is socially acceptable. Program action committees composed of community leaders can help. Group meetings, workshops, tours, and educational events are other ways in which evaluation can be fostered. It gives the individual an opportunity to get the feeling of others, as well as to get more information.

If the educator has helped the learner evaluate the information positively, the learner may be ready to begin a trial of the information.

Stage 4—Trial. At the fourth stage of learning, the learner puts the information into limited practice. They may try out a new variety of crops, another livestock enterprise plan, or a new and innovative piece of equipment.

The trial must be successful, and the educator must help the learner avoid failure. Personal help to ensure successful application at the time of trial is most helpful.

Final Stage 5—Adoption. Adoption of the information is dependent upon successful trial. If it works well for the learner, adoption may follow.

Continuing adoption is the result of successful trial, along with the continuing notice, assurance, and help of the educator. The educator must not let the learner fail.

Acceptance of Teacher/Change-Agents

Research supports the premise that development programs are more likely to be successful if change-agents are viewed by learners as persons whom they trust and respect. It is essential that the change-agent exhibit behavioral patterns which the learner can accept as appropriate in the culture and surroundings of the teaching site. There is evidence that these behavioral attributes apply equally to national change-agents as well as those from other countries.

Maintenance of Incentives to Foster Production and Develop Enthusiasm

If we subscribe to the theory that enthusiasm is the driving force behind production and development, then we must also recognize that success in achieving personal goals is the forerunner of enthusiasm. Where there are no recognizable successes, there will be no enthusiasm.

The following factors are crucial to agricultural programs, according to Bunch (1982, pp. 24–25):

1. the program must work toward solving felt needs (i.e., the people must want the problem to be solved).
2. the people must believe it possible for them to solve the problem (i.e., the solution must be simple and inexpensive enough to be within their means).
3. the people must believe that the program personnel know enough to be competent help, and are working for the people's benefit rather than to cheat or manipulate them.
4. the people should come to identify with the program's work and its successes by being involved in program planning.
5. the people must participate in the program's work so that when success is achieved, they will feel a sense of accomplishment. The challenges must be simple enough at first so that they can meaningfully participate, yet gradually become more complex so they can grow in their ability to deal with problems and can feel an increasing sense of accomplishment.

Maintenance of a High Level of Constructive Participation

Study of agricultural programs exhibiting high levels of appropriate technology-transfer and adaptation almost always reveal a common characteristic, that of widespread participation in certain essential decision-making and program-planning functions. Perhaps one of the most appealing definitions of development is that of a process whereby people learn to take charge of their own lives and solve their own problems.

Sociologists are quick to point out that the most productive participation patterns are those which are known as **Humanistic-Democratic-Participative (HDP) systems.** Contrasted with the more rigid traditional systems, the following characteristics of HDP practice are presented:*

- full and free communication, regardless of rank and power.
- a reliance on consensus, rather than on the more customary forms of coercion or compromise, to manage conflict.
- the idea that influence is based on technical competence and knowledge rather than on the vagaries of personal whims or prerogatives of power.
- an atmosphere that permits and encourages emotional expression as well as task-oriented acts.
- a basically human bias, one which accepts the inevitability of conflict between the organization and the individual, but which is willing to cope with and mediate the conflict on rational grounds.

In many countries, change and adoption occurring within the culture may well be accompanied by stress that is frustrating for potential adopters. Any change or attempt at technology-transfer, however simple that effort might appear to the citizen of a developed nation, appears quite demanding to many would-be adopters in developing countries. It would seem necessary that those who plan and implement change and encourage appropriate technology-transfer in other nations should work to bring about HDP management in as many organizational groups as possible.

The Outstanding Trainer/Change-Agent[†]

What separates the outstanding trainers from the mediocre ones? Three key concepts include the following:

1. base the level of instruction at the level of the group. You cannot communicate very effectively by talking down to trainees.
2. blend the instructional content with the backgrounds and experiences of the group. As the trainees continue to build new subject matter, skills, or attitudes from their own experiences, blending these items helps make learning much more effective.

---

*Reprinted by permission. Adapted from Jedlicka, *Organization for Rural Development.* Westport, CT: Prager Publishers, 1977, pp. 36–37.
†Adapted from Addison-Wesley Publishing Company, Reading, Massachusetts—Donaldson and Scannell, authors, 1986.

3. whenever possible, brighten sessions by using a combination of lecture, discussions, cases, and role play. Coupled with visual aids, an otherwise boring session can come alive.

In observation and discussions regarding the traits exhibited by outstanding trainers, trainees report common items. The traits most looked for in trainers include the following:

- knowledge of the subject matter.
- adaptability.
- sincerity.
- sense of humor.
- interest.
- understanding and willingness to involve the group.
- clear instruction.
- individual assistance.
- practicality of the subject matter.
- enthusiasm.
- other personal qualities such as promptness, neatness, friendliness, and courtesy.

Some items trainees have indicated they dislike in their trainers include the following:

- displays of superiority.
- lack of knowledge.
- unclear teaching.
- indifference.
- impatience.
- certain physical qualities such as a monotonous voice, a listless attitude, or a slovenly appearance.

If you are about to embark on an exciting and challenging international assignment, you are about to join the ranks of a special group of people known as human resource development professionals.

Conclusion

We would like to offer several proven points that can be used as tips to better training. These are not theoretical principles based on academic research; they are tested points proven through many years of experience in teaching. As you gain more experience in the dynamic field of international educating you will find your success may well depend on how these points are used.

1. Realize that trainees must want to learn. As mentioned in the earlier part of this chapter, there is only one way to get anyone to do anything—to make that person want to do it. As simple as this must sound, there is no other way! The learning climate must be conducive to good learning. Learning is largely a self-activity. By providing the proper motivational climate, you enable your participants to establish their own desires and motivation. By letting them know why a task or job is important, you are essentially telling your trainees they are important too.
2. The more we involve the use of the senses, the more we can learn. The hands-on or show-and-tell exercises are best for retention. As involvement increases, learning increases. On-the-job training is effective because of the real-world activity being learned through actual performance.
3. Adults will learn what they need to learn. Although you can force training, you cannot force learning. If a trainee does not desire or acknowledge the need to learn, it is all but imposssible to force that individual to learn. Give them the skills, attitudes, and knowledge they need and can put to use right now! Practicality is the key word and immediacy of application is paramount. While some sessions may rightfully teach the "nice to know," trainees will respond better to the things they "need to know."
4. Use problem-centered learning. Your trainees will learn faster and better if you use actual problems and let them work out the solutions. The discovery of knowledge or the formulation of principles through the case-study method is an excellent approach.
5. Experience connotes learning. Since we know that trainees learn from each other, it behooves us to capitalize on the backgrounds and experiences of our trainees. Exploit this wealth of talent and expertise in a positive way. Make use of the experience through discussion and group participation.
6. Keep things informal. It is a proven fact that trainees will learn best in an atmosphere of informality. Keep things informal, yet businesslike.
7. Artful use of visuals, coupled with a fast-paced presentation, makes for a good learning climate. A variety of methods and teaching techniques helps ensure a learning experience. Group participation is a very effective tool for learning; the lecture is perhaps the least effective.

Discussion Questions

1. Discuss learning as a process and as an activity.
2. Identify eight concepts as to how students (clients) learn.

3 Describe the characteristics which you deem desirable for a teacher or change-agent. What do you see as their major responsibility?
4 Discuss the stages of the adoption process.
5 Explain the four necessary attributes of educational programs designed to serve farmers.
6 Many studies have suggested that improved education results in development of a modern agriculture. Is this a good rationale for improving education in the rural sector of developing countries? Why or why not?
7 Explain why learning is an active process.
8 Explain the characteristics of the productive participation pattern known as Humanistic-Democratic-Participative (HDP) systems.
9 What separates the outstanding trainers from the mediocre ones?
10 What are some characteristics that trainees have indicated they dislike in their trainers?
11 Explain the first principle of learning. What implications does it have?

CHAPTER 13

Agricultural Education—Formal Instructional Services

Terms to Know
traditional society
future-oriented society
general education
vocational education
functional education
intellectual investment
enrollment ratio
pupil/teacher ratio
level of literacy
dual economies

Objectives

After reading this chapter, you should be able to:
- distinguish between vocational education and general education.
- list the three broad objectives of vocational education.
- describe the most common types of organizations at the various levels of education.
- describe the objectives of agricultural education programs at the various levels which are commonly found in developing countries.

- identify and discuss constraints and weaknesses often encountered at the higher education level in agricultural education programs.
- distinguish between a traditional society and a future-oriented society.
- understand why educational deficiencies lead to significant handicaps for any country.

The Nature of Society

In terms of the nature of desirable social and economic change, growth, and development, nations are sometimes classified into **traditional societies** and **future-oriented societies.**

Traditional societies are those in which productivity increases slowly. Traditional societies are primarily agricultural societies, since agriculture is the area of major activity for the population. The appropriation of goods and services produced from agricultural endeavors often constitutes the major source of wealth and power for the nation. The future-oriented society is one which constantly seeks to organize knowledge and skills to create and propagate commerce and industry as well as agriculture, and thus achieve an increase in the rate of discovery and application of new techniques.

A traditional society may become a future-oriented society as a result of achieving successive changes in social and economic structures. These changes are usually enhanced through effective educational programs. Education plays a decisive role in the emergence of a future-oriented society. It helps train research workers, instructs them in the use of scientific methods, and encourages using a creative approach. Furthermore, it makes the population at large more receptive to change and raises the general standard of information. Education is the foundation on which a future-oriented society can be constructed; however, the effectiveness of an educational system depends upon its objectives, methods, and structures.

General Education Compared to Vocational Education

In almost all countries of the world, society draws a distinction between **general education** (turning the children of upper-income families into "educated men and women") and **vocational education** (for the lower-income "ordinary people"). The prevalence of such a restricted philosophy and practice has at times slowed down the development process in agriculture.

Unfortunately, some developing nations bear the marks of Western educational models that separate general from vocational information. Training for agricultural occupations is thus provided in special establishments which issue their own diplomas, are seldom recognized as being equivalent to those of general education, and which do not permit access to other forms of higher education. Agricultural education is regarded, therefore, as an educational ghetto.

Vocational education is the part of education that makes an individual more employable in one group of occupations than in another. It may be differentiated from general education, which is of almost equal value regardless of the occupation that is to be followed.

There should be three basic objectives in any vocational education curriculum, and they are listed in chronological order of their acceptance as goals as follows:

1. meeting the manpower needs of society.
2. increasing the options available to each student.
3. serving as a motivating force to enhance all types of learning.

Functional education directed toward achievement of desirable agricultural change and development is, perhaps, one of the priorities that must be fully accepted by other nations. Education as a component of an agricultural system is sometimes referred to as an **intellectual investment.** Intellectual investment specific to the agricultural sector means expenditure devoted annually to agricultural research, instruction provided in agricultural schools and facilities, and agricultural advisory services. To these must be added farmers' organizations, cooperatives, and various forms of cultural activities, which are other forms of intellectual investment.

The Value of Education for Work*

According to Warnat (1991), vocational education in the U.S. has entered into a new era. The U.S. is beginning to value education for work. There is increasing concern that the U.S. is not adequately preparing the growing pool of women, minorities, and immigrants for productive roles in the work force, especially when compared to other countries such as England, Japan, Sweden, and Germany.

A U.S. General Accounting Office Study examined strategies used to prepare youth for employment in the U.S. compared to the previously mentioned countries. The findings include the following:

- all four foreign nations expect all students to do well in school, especially in the early years. U.S. schools routinely accept that most students will lag behind.
- the competitor nations have established competency-based national training standards that are used to certify competency. The U.S. practice is to certify program completion.
- all four foreign nations invest heavily in the education and training of youth. The U.S. invests less than half as much for each work-bound youth as it does for each college-bound youth.

---

*Adapted from the Vocational Education Journal.

- the schools and employment communities in the competitor countries guide students' transition from school to work, helping students learn about job requirements and assisting them in finding employment, much more so than in the U.S.
- young adults in the four competitor nations have higher literacy rates than the comparable youths in the U.S.

A study by the Education Writers Association compared the vocational education delivery systems of the U.S., Germany, Japan, and Korea. The quality of vocational education in Germany and Korea was considered to be uniformly good; in Japan, fair to good; in the U.S., from poor to excellent.

In terms of availability, vocational education is universally available in Germany and Korea, and of limited availability in Japan, as vocational preparation is mostly the responsibility of employers. In the U.S., availability is concentrated in more populated areas.

The global community requires a new kind of international relationship in which cooperation and coordination are of prime importance. Education, employment, and economic issues are central to this. It is imperative that we be aware of the similarities and differences that exist in vocational education everywhere.

The U.S. is unique in the following ways:

1. it is by far the largest country.
2. the U.S. operates vocational education through a highly diverse and decentralized system with 50 states providing the locus of control.
3. the U.S. system allows and encourages choice for all youth, provided they have adequate academic skills.

All advanced nations recognize that a higher level of education and training is needed by new workers. To compete in a global economy demands a literate work force, as well as one able to use higher-order thinking and problem-solving skills.

Organization of Agricultural Education in Other Nations

Agricultural education is conducted at many different educational levels. Although these educational levels are usually categorized in a hierarchical fashion, it is often more practical to categorize according to purpose and clientele. One possible classification is as follows:

1. the practical category, which deals with practical training of present and prospective farmers and other agricultural workers.

2. the primary and elementary category, which provides basic knowledge about agriculture for in-school youth.
3. the pre-vocational category, which is preparatory to beginning vocational training or entering agricultural vocations.
4. the vocational category, designed to prepare students for farming, is carried-out in agricultural high schools or comprehensive high schools.
5. the sub-professional or technical category, for preparation of technicians in government agricultural services.
6. the professional category, largely provided through the university or college.

Recent developments point to a tendency to do away with primary instruction in agriculture and a tendency to raise the sub-professional category to a degree course by adding one or two years to existing junior college and pre-university curricula.

In many nations, the administration of formal agricultural education at all levels comes directly under the supervision of the Ministry of Education or the Ministry of Agriculture. However, in some countries, the operation of the different categories of agricultural education is carried out independently by both ministries concerned. Programs at the practical levels designed for present and prospective farmers are the responsibility of the Ministry of Agriculture or a Technical Ministry. Programs and objectives of agricultural schools under each ministry are quite similar, except that the Ministry of Agriculture directs training more specifically toward preparation of graduates for eventual employment in the Ministry. Consequently, such training programs can be considered as terminal. Graduates from such programs, if they desire further formal education, are often forced to make up for certain alleged academic deficiencies in order to meet existing college or university requirements.

The quality and quantity of facilities in many of the agricultural institutions vary from quite adequate to virtually nonexistent. Facilities often reported as being inadequate include laboratory supplies and equipment, farm lands, farm tools and equipment, textbook reference and other library materials of local importance, and classroom and laboratory space. The prevalence of western publications has been noted in the libraries of some Asian countries, but these have little application to local conditions of climate and soil.

The lack of substantial incentive or reward for workers in agriculture is also reflected in the meager financing of agricultural schools.

The greatest constraint to narrowing the gap between food produced and the amount needed is lack of incentive. Producers simply are not provided motivation to take the risks involved in making needed changes or adopting new, improved practices. An integral part of this constraint is the prevalence of antiquated knowledge delivery systems. This lack of communication points directly to the efficacy of educational systems as an essential component of any plan directed toward improved agricultural production and food distribution.

Primary, Elementary, and Secondary Schools of Developing Nations

Progress toward improving literacy rates in developing nations has often been sporadic. While in some countries the percentage of classified illiterates has slowly dropped over the past two decades, this statistic has remained essentially unchanged. It is estimated that more than 375 million children aged 7 to 12 are not currently attending school.

Even those fortunate enough to be in school must overcome the effects of a vicious cycle. The malnutrition accompanying dire poverty leads to limited mental and physical development which, in turn, leads to school failure, poor work opportunities, and continued poverty. The cycle begins again when another generation is born. Recent research has verified that the poorer the country is, the less children learn in schools. Heyneman (World Bank Reports, 1983) reported that teachers, furniture, equipment, and materials in schools of many developing countries are well below the standard considered minimal in the industrialized societies. The value of furniture and materials in the average fourth-grade classroom in Bolivia is 1 percent of that in U.S. classrooms. Furthermore, the gap in classroom quality between high- and low-income countries is widening; as more pupils enter school in the developing countries, there are less materials available with which to teach them.

Such educational deficiencies lead to significant handicaps for any country, particularly since these deficiencies can be undeniably associated with retarded economic growth, inefficiencies in agricultural productivity, and a higher rate of breakdown in the maintenance of physical plants. A failure in education inevitably leads to less effective investments in health, population control, nutrition, and specialized skill training.

Differences in educational resources available to elementary pupils in certain industrialized nations as compared to some other countries are emphasized as follows:

- 33 percent of elementary pupils in the U.S. have access to a computer
- 97 percent of the elementary schools in Japan have a tape recorder
- 27 percent have a color TV camera
- virtually all schools in Japan have an overhead projector, a slide projector, and an 8 mm projector.

There are 71,000 school libraries in the U.S. with an average of 14 titles for every student, in addition to textbooks, reference books, and visual aids. Compare this with the Philippines, where there are approximately 10 pupils for each available textbook. The typical pupil in the U.S. has 140 times the amount of reading material at his disposal for accomplishing similar objectives—the ability to calculate, read, write, and comprehend the world around him.

Let us examine five African countries: Chad, Malawi, Zimbabwe, Zaire, and Cameroon. Based on the UNESCO STATISTICAL YEARBOOK 1987 as quoted by

Table 13–1 •

Country	Gross Primary Enrollment Ratios (%)
Chad	35
Malawi	61
Zimbabwe	88
Zaire	98
Cameroon	104

Meaders, et al. (1988),* of particular interest are enrollment ratios, pupil/teacher ratios, and levels of literacy.

The **enrollment ratio** is the ratio of enrollment at a given level of schooling or at a given age group to the relevant population. The gross enrollment ratios for primary education in the five selected African countries, based on the UNESCO report, show a great disparity in primary school capacity. The enrollment ratios for primary education in 1980 ranged from 35 in Chad to 104 in Cameroon, as shown in Table 13–1.

The ratio may surpass 100 because of the inclusion of students who are under- or overage in relation to the designated age group.

The **pupil/teacher ratio** shows the class size. The pupil/teacher ratios for the five African countries were reported to be as follows:

1. Cameroon (1984) 51:1
2. Chad (1984) 64:1
3. Malawi (1983) 60:1
4. Zaire (1984) 42:1
5. Zimbabwe (1984) 39:1

When compared with the pupil/teacher ratios in developed countries, this indicates a rather large class size. The pupil/teacher ratios for the same year (1984) were 17:1 in Canada, 21:1 in the U.S., and 24:1 in Japan (UNESCO, 1987).

The general **level of literacy** within a country is believed to affect the achievement of development goals. High levels of illiteracy tend to inhibit development. For agricultural development it is important to analyze the illiteracy figures. The illiteracy rates (1980) for females and males in the five countries have been reported as follows (Cisse, 1986) in Table 13–2.

There is a wide range in the levels of illiteracy between females in Chad and males in Zimbabwe. Given the importance of women in agricultural development, there are wide disparities between illiteracy of females and males.

---

*Reprinted by permission of O. Donald Meaders.

Table 13-2 • Illiteracy Rates

	Females	Males
Chad	99.5	63.5
Malawi	80.7	53.4
Zaire	60.6	22.8
Cameroon	59.0	29.8
Zimbabwe	36.2	22.0

When females receive an education, the benefits are multiplied. The better educated the mother, the less likely the child is to die in infancy. Studies from developing countries show that four to six years of education is associated with a 20 percent drop in infant deaths. The children of better educated mothers are better nourished, healthier, and more likely to succeed in school. An educated woman is more productive at home and in the workplace, and better positioned to get further education.

Educated women are more receptive to family planning and tend to have later marriages with fewer children. A study of four Latin American countries discovered that education was responsible for 40 to 60 percent of the last decade's fertility decline.

Education also empowers women to exercise their rights and responsibilities as citizens. But female education is at a lower level than that for males in most developing nations. In 14 of the developing countries where literacy data are available, only one in five adult females can read. In some of the poorest countries, only 5 percent of women are literate. Only one out of two women in Asia is literate and only one out of three in sub-Saharan Africa.

Despite the benefits of an education, there are real constraints to access to schools for females, and these constraints are economically based; education costs money. Sending daughters to school also means a loss of work by daughters at home, in the field, or at the marketplace. And there are cultural factors as well—education is not always an accepted social norm for females.

Agricultural Education in Paraguay: An Example of Success*

There have been many successful implementations of agricultural education programs around the world. Paraguay serves as one example as reported by Welton, et al. (1987).

---

*Adapted from the National Association of Colleges and Teachers of Agriculture [NACTA] Journal.

Paraguay is a landlocked South American country that will soon become the largest exporter of hydroelectric power in the world. Currently, the economy of Paraguay depends basically on agriculture; however, since the completion of Itaipu Dam, the exportation of hydroelectric power should contribute an increasing share of the economy. Currently, 36 percent of the gross national product, 85 percent of the value of exports, and 51 percent of the nation's employment are provided by the agriculture sector. The training of farmers is one of the highest priority programs of agricultural education in Paraguay.

During the 1970s, the Ministry of Agriculture and Livestock, through the Directorate of Agricultural Education, initiated an ambitious plan to build a network of agricultural high schools throughout the country. This effort was supported by a loan from the International Development Bank. The Bank sponsored two projects to expand and strengthen agricultural education. Three of the four national schools and one private school were modernized under the first program. The second project provided three new national schools, with one of the old national schools as well as a private school being remodeled. Training of personnel from these schools took place in Paraguay and other countries.

At the present time, there are 19 agricultural schools located throughout Paraguay. All schools are designed to board students. They also have facilities needed for instruction in technical education in agriculture and each has its own farm; crops and livestock produced on these farms are representative of conditions of the local community. School farm products provide a source of revenue for the school as well as part of the food for the cafeteria.

The principal purpose of the curriculum in technical agriculture is to train technicians for the occupational needs of the agricultural sector. Both curriculums are more practical than theoretical and students learn by doing on the farm.

Relationship of Instruction at Primary and Secondary Levels to Agriculture in Developing Countries

Although many nations have in the past required that agriculture be taught at the primary or elementary level, such instruction has only rarely been supplemented with any practical application. Rote memorization of definitions of agricultural terms often occupies a considerable portion of time allocated for study. At the secondary level, even though programs may be designated as vocational agriculture, opportunities for supervised work experience in agriculture are often quite limited, even in instances where a school farm may be in operation. The passing of an examination in order to gain access to a higher level of study tends to stultify any efforts toward acquisition of knowledge and skills through learning by doing. While we cannot, and should not, attempt to transfer whole systems of education from industrialized nations to developing nations with highly traditional educational patterns and practices, a few principles can be suggested as worthy of consideration for effecting improvement and are as follows:

1. raise the prestige of agriculturally related studies.
2. attempt drastic changes in the requirements for gaining entry to higher levels of study.
3. attempt to provide work experiences in agriculture for secondary school students.
4. consider the values which may accrue when greater resources are allocated to the educational sector.

Institutions of Higher Education in Developing Countries

It has been generally accepted in the U.S. that instruction, research, and extension should be the triple functions of institutions of higher education in agriculture. In the U.S., the Morrill Act, providing for the Land Grant University System, specifically pointed to instruction. The functions of research and extension were officially added later. However, among the developing nations only a very few have adopted this concept, particularly with regard to administrative structures. It must be recognized that agricultural education extends all the way from farmers, to those who serve farmers directly and indirectly, to those who form the policies that affect farmers. Institutions of higher education in agriculture must take these facts into account.

Institutions of higher education in developing nations constitute one of the most important resources for promoting and enhancing agricultural production and development. Often they are the repositories of talent in which substantial investments have been made. However, their effectiveness in serving agriculture is frequently impaired by weaknesses such as the following:

1. little or no involvement of the college or school in the nation's efforts to substantially improve agricultural production or rural development. The potential role of the institutions in agriculture is either not recognized or is assigned a low priority.
2. college or school experiment stations, even when functioning in the field of agriculture, are often ill-maintained and underutilized. Rarely is there any attempt to provide experimentation focused on the farm level.
3. perhaps as a result of being structured under a ministry other than agriculture, the college or school may be functioning at a level essentially out of touch with the mainstream of the nation's agricultural industry.
4. faculty in agriculture, although perhaps academically able, often lack agricultural skills or field experience. Consequently, they tend to confine themselves to classroom teaching or to research based largely upon literature or upon

laboratory work which is merely an extension of studies initiated in graduate school. Such instructors are ill-equipped to transmit skills to students via joint involvement in field experimentation. The all-too-common lack of dynamic, purposeful, and applicable field research at institutions offering higher level studies in agriculture could well be the result of a lower level of competence—and confidence—among the faculty to undertake such endeavors.

5. the students in attendance at institutions of higher education in developing nations are usually from urban areas. Even rural students are generally from subsistence farms and do not have experience in field management of crops or animals for high productivity. Unfortunately, they do not obtain this experience at the college or school. A student may graduate with an education but with very few accompanying skills. Often the student is then faced with frustrating situations in assuming the responsibilities of a change-agent without having developed confidence in facing farmers and real farm problems.

6. an additional weakness related to number five above is inherent in an attitude that quite often tends to perpetuate, rather than dissipate the philosophy that working with the hands, or manual labor, is beneath the dignity of the truly educated person. To engage in activities that might involve field contact with the soil or with animals is, in many cultures, considered inappropriate for the educator.

Involvement of U.S. Universities

Title XII of the Foreign Assistance Act of 1975 provided for the entitlement of universities and colleges of agriculture in the U.S. to render assistance to their counterparts in Third World countries in agricultural education, research, and development. Among a number of provisions, the legislation called for expenditures of funds to provide faculty interchange and to encourage visits by international students' advisors to the students' home countries to assist in gathering research data. The National Association of State Universities and Land Grant Colleges asserted that such participation strengthens both international and U.S. higher education in agriculture. The National Association of State Universities and Land Grant Colleges enumerated further the following propositions (reprinted by permission):

Educated and trained people at all levels provide the key to economic and agricultural development in other countries. The need still exists for more and higher quality national schools, universities, and research centers to train these people and to conduct problem-solving research. U.S. universities and colleges of agriculture can be instrumental in the development of such institutions. Ingredients essential for successful university involvement in international programs include the following:

- full commitment of interest and support of the university administration, including deans, department heads, and faculty to the international agricultural dimension.

- integration of international activities into research, teaching, and extension functions at the departmental level.
- long-term commitments of universities and their resources to meet the challenges projected for international agricultural development.
- long-term financial support for overseas institution building and agricultural development projects primarily from federal, foundation, or international agency sources. Additionally, states have the responsibility to provide the resources needed by their universities for the on-campus education and training of students they accept. States further have the responsibility to support research that will benefit domestic programs—research that may have to be conducted in other countries.
- commitment by American universities and colleges of agriculture to develop the capacity, support, and resources to serve as contributing and receiving partners in the international research and training network.

For additional reading, refer to Appendix G and review the article entitled, "Basic Principles for College and University Involvement in International Institutions." Also, refer to Appendix E and review the comments presented by Schuh (1987) regarding "The Role of the Land Grant University: An International Perspective."

Development Education for Citizens of Developed Countries

While enlightened citizens of developed nations may know something of the geography of other countries and even have some knowledge of their history, culture, and customs, they are often painfully ignorant concerning the welfare of citizens of other countries in such areas as health, nutrition, education, and economics. Even studies of governmental functions may be somewhat distorted. This lack of knowledge about the true character of people in the poorer, low-industrialized nations can carry an implication that the poverty of such nations is a consequence of laziness and apathy on the part of their citizens, who have little or no desire for education or for improved living conditions.

The United Nations organization, largely through UNESCO, is exerting much energy to promote truly effective development education for children and youth in developed nations. Although this approach seems largely to be directed toward elementary and secondary levels, the need for implementing development education into curricula in higher education would seem to be of equal importance. The promotion of development education to create an awareness and understanding of true needs is a task being carried out with reasonable success by the media and by organizations dedicated to such a purpose.

Similarly, other countries maintain institutes and training centers more specifically designed to prepare professional change-agents such as adult educators, local development center directors, researchers, and extension personnel.

Geographic Education for U.S. Citizens

In 1988, the National Geographic Society's centennial year, the Society commissioned an international Gallup survey to test geographic literacy. The countries tested were the U.S., Sweden, West Germany, Japan, France, Canada, the United Kingdom, Italy, and Mexico.

Americans ranked in the bottom third. One in seven people could not identify the U.S. on a world map and one in four could not find the Pacific Ocean. Fewer than half could pick out countries often mentioned in the news such as the United Kingdom, France, South Africa, and Japan. The results were worst among young Americans between the ages of 18 and 24 (National Geographic Society, 1990).

In order to understand global economics, population and hunger problems, arms control, and the need for international care of the environment, it is essential that we all come to know the world.

International Extension Programs For U.S. Citizens

Refer to Appendix D and read Schuh's (1985) comments regarding "international extension programs for U.S. citizens." Schuh discusses the changes in the international economy, the emergence of a well-integrated international capital market, the shift to a system of bloc-floating currency exchange rates, increased monetary instability, and then makes recommendations regarding the content of international education programs for U.S. citizens.

Influences of Society on Education

In all countries there are many influences of society on education. Before we examine these influences, let us examine some of the characteristics of society within many of the developing countries. Many developing countries are made up of a society which exhibits the following characteristics:

- **dual economies,** the rich are very rich and the poor are very poor, as compared to the U.S. where a large proportion of the population is composed of middle-income families.
- underutilization of resources.
- absence of corrective institutions.
- difficulties in communications.
- unwillingness to trust individuals outside the family.
- aversion to risk.
- high preference of some classes for leisure.
- cultural inconsistencies.
- citizens having little access to the political administration.

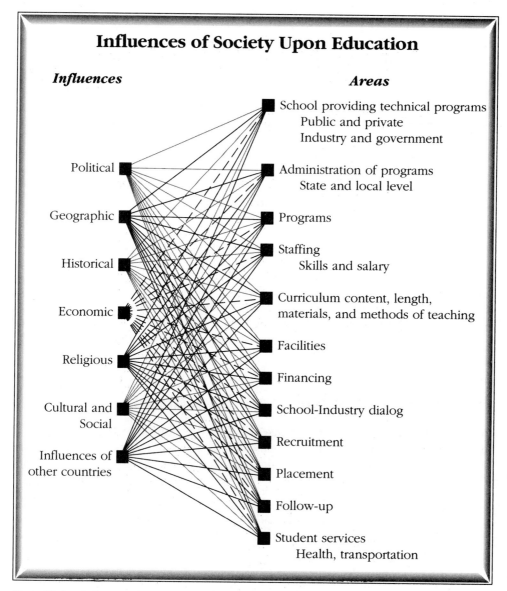

Figure 13–1 • Influences of society upon education.

- not everyone treated equally under the law.
- no means by which political control is transferred effectively.

Figure 13–1 graphically depicts the influences of society upon education. All areas of education are influenced by politics, geography, history, economic conditions, religion, cultural and social acceptances, and influences of other countries.

Conclusion

A nation's success in expanding agricultural production will depend in large part on its ability to produce substantial numbers of people who exhibit the following:

- understanding of national goals and objectives.
- command of basic farming skills.
- good background in agricultural science and technology.
- that they are oriented to acting rapidly.
- ability to direct and participate in commodity production programs and defined area campaigns.
- capability of drawing public and private institutions into development activities in a concerted way.

Most countries already have a variety of institutions that prepare people for agricultural careers. Strong training in academic disciplines such as crop and animal sciences, basic chemistry and physics, economics, and sociology is highly desirable for agricultural college graduates. Also, there is a great need for graduates who have expertise in extension teaching methods and agricultural technology-transfer techniques.

Training should ensure that each individual has opportunities and motivation to learn in three interrelated areas. First, every agricultural worker should have basic farming skills. Second, each agricultural worker should acquire the specialized knowledge necessary to make them effective at their assigned duties. Third, each agricultural worker should have management competency to enable them to organize their own efforts and, in appropriate situations, to assist in organizing commodity production programs, defined area campaigns, or other problem-centered activities.

The developing countries have too few agricultural professionals who understand farming. One reason is that a higher proportion of students finish secondary school in urban areas than in rural areas. More urban students then go on to colleges and universities where they obtain agricultural degrees. They have not grown up on farms.

Many countries must find a way to break a system which perpetuates poor-quality agricultural training. The U.S. experience is instructive for understanding how one country educated an agriculturally literate population as well as a diverse array of competent scientists, administrators, teachers, technicians, businessmen, and farmers with a highly developed service orientation to agriculture.

Finally, encouraging increases in world literacy have been evident in this century. In 1900, some 90 percent of the world population could not read or write. The number today is approximately 50 percent. Much of today's illiteracy is in the developing world, where skills are badly needed. Agriculturalists and teachers need to work hand in hand because the quest for food relates to a people's ability to read and write.

Discussion Questions

1. Define a traditional society. How does it differ from a future-oriented society?
2. List and describe the various levels of education and explain the two major trends related to these levels.
3. Why is "intellectual investment" so essential to a nation's agricultural development?
4. Which situations most commonly tend to restrict the effectiveness of agricultural education programs?
5. What can be said concerning the need for development education among citizens of developed countries?
6. Why should we not attempt to transfer whole systems of education from industrialized nations to those nations with limited resources?
7. Explain why institutions of higher education in developing nations constitute one of the most important resources for promoting and enhancing agricultural production and management.
8. What did Title XII of the Foreign Assistance Act of 1975 provide for?
9. Briefly describe the influences of society on education. Are the influences different in the U.S. as compared to other countries? What are some of the differences/similarities?
10. What are the three broad objectives for vocational education?
11. Why is education for females not considered to be a social/cultural norm in some countries?

CHAPTER 14

Agricultural Extension—Providing Instructional Services

Source: Food and Agriculture Organization; Photo by I. Bara.

Objectives

After reading this chapter, you should be able to:

- discuss the necessary requirements to develop and maintain an effective agricultural extension service.
- briefly describe the organization and function of the Cooperative Agricultural Extension Service in the U.S.
- analyze the strengths and weaknesses of typical agricultural extension services in developing nations.
- identify the resources needed to establish and maintain an effective agricultural extension service.
- explain the importance of careful selection and adaptation of extension teaching methods.
- better understand the role of women and rural youth in production agriculture.

Terms to Know

Cooperative Extension Service (CES)
Smith-Lever Act
cooperators
Ministry of Agriculture
microentrepreneurs
rural young people
ancillary services
human resources
physical resources
scientific knowledge
financial resources
administrative and political support
communication media
result demonstration

The Importance of Agricultural Development and Training

During a World Food Conference in Rome, Italy, it was determined that priority be given to the development of agricultural education and training, including training of research and extension workers in management techniques, special basic and inservice training for graduate and middle-level extension personnel, and farmers' training programs for rural women and children. The aim was for the achievement of an integrated educational system for the rural population within an appropriate political and social framework.

The government of each nation must be responsible for the production, processing, and distribution of food for its own people. Only the individual nation can determine the relative amounts of foodstuffs to be imported or to be grown locally, and only the government can set the policies, establish the priorities, allocate the resources, and involve the farmers.

Universities can play an important role in improving the services provided by ministries of agriculture and agricultural extension services. These institutions should become actively involved with other agencies to strengthen extension work. Our American universities represent a large pool of skilled professsionals who can assist the universities abroad. The universities abroad must be the leaders in the task of providing for increased agricultural production and income.

Extension programs can have a tremendous impact on the agricultural and rural development of every nation. The economic development of these countries depends to a large extent on their capacity to develop and improve their agricultural resources. Governments could assist in the agricultural development of their nation by establishing and maintaining a well-trained group of extension agents to carry the results of agricultural research to the farmers. Assisting emerging nations to develop an effective extension program may be the greatest contribution that developed nations can make.

The extension program of a nation should be organized on a legal basis. This would make continuing financial support and recognition by the government more likely. Administrative and policy-making relationships of extension with other governmental agencies should be curtailed or severed. The legal basis for extension should be relevant to each country's political structure and should provide for adequate financing and staff to meet the needs of the smallest political entity. Basically, extension contacts the individual farmer and teaches him how to manage his own resources (land, labor, and capital) efficiently. Figure 14–1 depicts the extension linkage with research and farmers.

Figure 14–2 depicts the linkages of supportive services which farm families require in order to sustain and improve their quality of life, production, and marketing.

Governments should give high priority to food production, processing, and distribution. Extension service workers should be adequately trained and placed in agricultural areas in sufficient numbers to serve producers effectively, with special emphasis given to repeated contacts. Efforts should be made to identify and select the most promising future extension leaders. A number of these should be sent abroad to study successful extension operations in other nations. Governments should expand their agricultural production and should adequately address the necessity of an effective educational system.

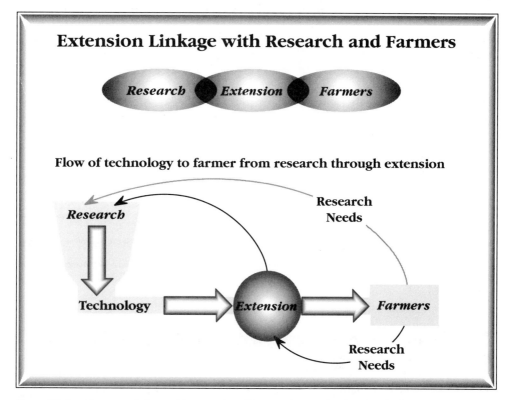

Figure 14–1 • Extension linkage with research and farmers. *Source: Food and Agriculture Organization of the United Nations, Rome, Italy.*

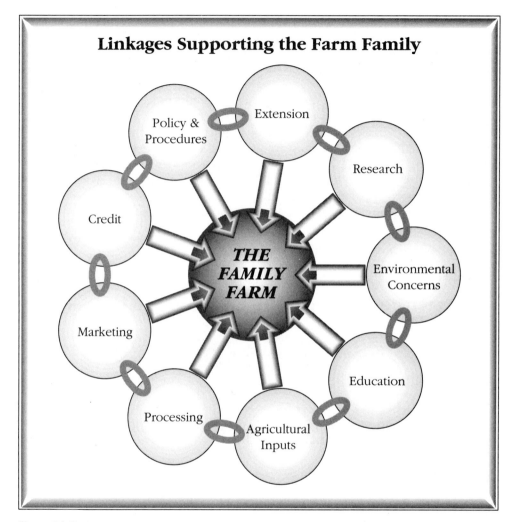

Figure 14–2 • Linkages supporting the farm family. *Source: Food and Agriculture Organization of the United Nations, Rome, Italy.*

The Cooperative Agricultural Extension Service in the U.S.

The **Cooperative Extension Service (CES)** is often given credit for the phenomenal success of agricultural producers in the U.S. Developed as an integral part of the Land Grant System, originating from national legislation known as the Morrill Act, Cooperative

Extension is sometimes referred to as the agricultural arm of the university. A description as given by Ebling (adapted from the University of California Press) is as follows:

Cooperative Extension is an undertaking of the USDA, the state land-grant institutions, and the farmers, through the provisions of the **Smith-Lever Act.** The cost is borne by the federal government, the states, and the counties.

The Smith-Lever Act provides federal aid for agricultural extension work. For a state to obtain its share of federal funds each year, it must submit plans for agricultural extension programs to the Secretary of Agriculture and secure approval in advance, and must provide matching funds. Counties also cooperate by contributing funds. Each county has what is called a county director and farm adviser or home adviser. The county cooperative extension agent can be called upon for information or counsel. County educational programs include farm calls, meetings, newsletters, and use of mass media. The county extension agent utilizes resources of the county, state, and federal governments, but is usually an employee of the state agricultural college, conducting off-campus, noncredit, teaching programs. There are also extension specialists, such as extension agronomists, entomologists, economists, horticulturists, and foresters, to assist in bringing research results from the college to the extension agents. The federal extension service also carries the results of Department of Agriculture research to the state agricultural colleges, thereby incorporating them into the extension program of the state.

The fact that farmers are also experimenters and innovators deserves emphasis. They often permit the county agent to develop test plots and conduct experiments on their land to develop results for the benefit of producers in the area. These farmers are termed **cooperators,** working with the university in developing and carrying out plans, each supplementing the knowledge and experience of the other.

Extension is Educational by Design

Extension is educational, but differs greatly from the common concept of an educational institution in that it does not have a fixed curriculum or course of study. It operates informally off-campus and uses farms, homes, and businesses as classrooms. Furthermore, the extension agent has a large field of practical subject matter that is intended for immediate application in the solution of problems. Participation in extension programs by farmers and others is purely voluntary. Extension has the freedom to develop programs locally that are based on the needs and expressed desires of the people; however, these programs are based largely on research. Furthermore, extension programs are designed in order to be of assistance to all members of a family.

Organization of International Agricultural Extension Programs

Almost all agricultural extension programs operate under the administration of a **Ministry of Agriculture.** While their responsibilities vary to some extent, they are often in charge of production programs and agricultural credit farm cooperatives. In some instances they are also involved in certain phases of marketing. One disadvantage of agricultural extension services in developing countries is that they often are burdened with a somewhat autocratic bureaucracy and generally are proportionately overstaffed at the top levels of the hierarchy.

Areas In Which Improvement is Needed

Even though agricultural extension services in many nations do contribute significantly to advances in food production, a number of weaknesses can be identified. For example, a group of delegates from 17 East African nations identified certain extension problems.

The main bottlenecks in extension work mentioned by the majority of the extension staff in East Africa were as follows:

- lack of qualified staff.
- lack of marketing facilities which offer fair prices for agricultural produce.
- lack of transport and travel funds.
- lack or delay of communications (both within and outside).
- lack of credit for farmers.
- lack of channels for agricultural supplies.
- too many transfers of extension personnel.
- too much dependence on foreign aid.
- lack of understanding of extension concepts.
- too high a proportion of farmers per field worker.

Further evidence available indicates that there were other problems, as follows:

- lack of clearly defined extension objectives at all levels, lack of job descriptions for extension personnel and lack of sufficient personnel and program policies.
- lack of annual extension plan, lack of evaluation activities, and lack of extension manuals for personnel.
- lack of information on suitable level of training required by extension staff.
- lack of economic and farm management research for dissemination.
- in most instances, economic planners did not have any idea regarding the role, importance, and problems of extension work. Field extension workers often were not acquainted with their countries' development plan.
- lack of two-way communications between research and extension, lack of flow of research information to extension, and lack of farmers' problems being transmitted to research institutes.
- lack of extension supervisors especially trained for the job.

- lack of supporting services or lack of coordination between these (especially credit facilities) and extension services.

The Role of Extension Field Workers

There is growing recognition that increased food production will come from investment in training and providing incentives for the extension field worker. Extension workers are the conduit for agricultural technology, technical knowledge and skills, agricultural credit, organizational services, land reform, and general development information. There are those who argue that if the technology is sound and profitable, it will diffuse without the aid of extension workers. But where farmers are poor and have minimal education, diffusion and adoption cannot take place rapidly without the necessary knowledge to use technology correctly. The extension worker is the farmer's link with the outside world, and often determines which farmers will receive new seeds, credit, and fertilizer. Where the farmer/extension worker ratio is very unfavorable, the extension worker cannot give priority to one sector of the farming community over another.

At the local level, the role of the extension field worker must be directed toward training farmers as adopter/leaders or promoters. This enables the field worker to service people in many more communities, but still provide a wide range of up-to-date services; see Figures 14–3, 14–4, and 14–5.

Women as Farmers and Farm Laborers*

A significant proportion of small farmers and farm workers in many countries are women. Estimates of women's contribution vary widely, but all estimates suggest that women constitute a significant proportion. Women's active roles in subsistence farming and in production for market exchange have been well documented, and their input in agriculturally related decision-making is substantial; see Figure 14–7. Empirically based case studies in South America, Africa, and India suggest women's input varies among geographical regions and in relation to farm size. These studies tentatively conclude that women influence men's decisions in specified areas, such as the purchase and use of certain modern inputs and the joining of cooperatives. For women who work along with their husbands, it is assumed that extension efforts reach them through their husbands.

U.S. AID Highlights (1991) presented the following example of the importance of women in agricultural production (Reprinted by permission of the United States Agency for International Development):

---

*Adapted from the Food and Agriculture Organization of the United Nations, Rome, Italy.

Figure 14–3 ● Bangladesh has launched its first community forestry project. Broadly speaking, the project involves setting up, with the assistance of UNDP and FAO, appropriate training programs to give present and future staff the necessary competence and skills. But it also involves training the trainers—that is, making sure the trainers are not only competent in technical knowledge of forestry, but that they know how to effectively pass this information on to others. *Source: Food and Agriculture Organization of the the United Nations; photo by I. Bara.*

Boonlieng Lattisoongenin is a member of the committee responsible for the management of a feed mill in Thailand's Saraburi Providence. The mill was set up as part of a USAID-supported project to involve women in agriculture. As a result, Lattisoongenin was able to more than double her number of cows, buy new milking equipment, and upgrade pasture land. Today, she is saving more money than she used to earn in a year and using some of her savings to send her fourth child, a girl, to the university. "My daughter is majoring in business so that she can take over my work," she says.

The contribution of women like Boonlieng Lattisoongenin to the welfare of families and the wealth of nations is increasingly recognized.

Women are critical to economic development. They are the sole breadwinners of one out of every three households in the developing world. They

Figure 14–4 • FAO Associate Professional Officer Training in Burundi. An Associate Professional Officers Program is offered by the Food and Agriculture Organization of the United Nations. It provides valuable training to young professionals in the field of agricultural development. Pictured is an Associate Professional Training Officer inspecting the Murambe water system in Burundi. *Source: Food and Agriculture Organization of the United Nations; photo by C. Reynolds.*

are farmers-producers of more than half of the world's food. They are **microentrepreneurs.** And they play the primary role in family health, nutrition, and education. Yet women have only limited control over income and limited access to credit, land, education, and training. They receive only one percent of the world's income and own less than one percent of the world's property.

"We strongly support greater efforts to integrate women into the development process," President Bush told participants at the annual meeting of the International Monetary Fund and the World Bank. "The key to economic growth is setting individuals free—free to take risks, free to make choices, free to use their initiative and their abilities in the market place."

Figure 14–5 • Fire-fighter trainees put out a bush fire with water during a drill near Chandrapur, India. FAO experts are assisting training of locally hired fire-fighters at two demonstration sites (flat and hilly). The project, a pioneering effort and unique in the Third World, is also promoting fire-prevention awareness among villagers. *Source: Food and Agriculture Organization of the United Nations; photo by I. de Borhegyi.*

Women contribute to family welfare. Their income is used to provide food and schooling for their children. Edite Genegani, for example, is a Filipino housewife who works packaging mushrooms in a rural area outside of Manila. Working part-time, she earns 200 pesos ($10) a week, about two-thirds of her husband's full-time salary. "I am saving my money for the education of my children," she says. "The only wealth we poor people can give our children is education."

Growth in women's income leads directly to improvements in family health and education. Women are also good credit risks. They pay back loans. More than 80 percent of 9,000 loans made to rural women in Bangladesh through a USAID-supported small business project were paid back on time and in full.

Women comprise more than 40 percent of the world's agricultural workers—a number that is growing as more men in developing countries migrate to cities and towns to seek employment, leaving women behind to work the land.

Figure 14–6 • Women's work is never done.

Women are involved in every type of agricultural activity, putting in long hours as producers, distributors, processors, storers, and marketers of food. More than half of the world's food is produced by women. And in many parts of the world, raising poultry and looking after livestock are tasks often left exclusively to women.

Despite their central role in agriculture, women do not have the same

Figure 14–7 • Like many African women, this woman in Sierra Leone juggles the responsibilities of wife, mother and farmer, often keeping an eye on her children as she works in the fields. Some 80 percent of the food produced in Africa is grown by women. *Source: CARE; photo by Vera Viditz-Ward.*

access as men to training, extension services, technology, credit, and ownership of land. Various cultural, traditional, legal, and policy barriers exist to deny them the same flexibility and productivity enjoyed by male farmers. As a result, women's agricultural productivity is lower than it should be—a fact reflected in the low national agricultural output in many nations.

For women, especially in the developing world, a healthy environment is the foundation for survival. The ability of women to produce food, to generate income, and to provide fuel and shelter for their families is closely linked to the well-being of the earth's natural resources.

As farmers and providers of household water and fuelwood supplies, women have a direct stake in the conservation and sustainable management of natural resources. Deforestation means have to spend more hours collecting wood for fuel. As less wood is available, treks to forage for fuel become longer. Women often must walk more than five miles and spend five to eight hours every day collecting wood for fuel. Figure 14–6 depicts a day in the life of a typical rural African woman.

Facts About Women in the Developing World

Women comprise an estimated 32 percent of the measured labor force in developing countries. Roughly one-third of all households in the developing countries are headed by women. Poor women seldom stop working. They work longer hours than men. Discrimination against girls is widespread. Eight out of nine cultures that express a preference want more sons than daughters.

Facts About Women Who Farm in the U.S.

The traditional image of the farmer in the U.S. has been masculine, but women always have played a major role.

In the past, most women on farms were not viewed as being in the labor force, regardless of the farmwork they may have performed. A minority were counted as unpaid family workers. Only a farm widow was likely to be viewed as a farm operator. Over the years, more women have entered farming on their own, and farm women spouses have more frequently come to view themselves as cooperators with their husbands. At the same time, so many farm women have taken off-farm jobs that nearly three fourths of those who now work do so primarily in a nonfarm industry.

The 1987 Census of Agriculture identified 132,000 farms whose operators or senior partners were women. This represented 6 percent of all farms, and was an increase of 10,000 in five years, at a time when the overall number of farms was decreasing. A large minority of these women are widows, for nearly one fourth of them were 70 years old or over, twice the percentage found among male farmers. Nevertheless, there are many young women who are in farming.

Women in the U.S. are involved in all types of agriculture. Their most common enterprise is beef cattle, in which more than one-third of them specialize. Their greatest relative presence, though, is in animal specialty farming, where they own one sixth of the farms. The greatest number of these are in horse-breeding operations. Poultry, horticultural production, fruits, and nuts are other types of farming favored by U.S. women farmers.

Rural Young People*

Rural young people is another group that has received too little attention in agricultural extension programs. The millions of young people living in rural areas are a significant and untapped resource.

The United Nations defines rural young people as individuals from 15 to 24 years of age. It is estimated that 20 percent of the world's population falls in this category, and there will soon be more than 742 million young people living in less developed countries. Of this, over 70 percent live in rural areas, and the majority are victims of rural poverty.

The opportunities for formal education and technical training for youth living in rural areas are very limited in many countries, especially for girls. The number of school drop-outs is high, and too many rural children never attend school.

The consequences of this situation is particularly serious. These young people tend to have a very low level of functional literacy.

Rural youth has a widespread need for practical training in agriculture, home economics, group leadership, and progressive rural living. Nonformal education programs provide some opportunities for training, but the low ratio of extension workers to farm families, and the general lack of preparation to organize and support youth programs severely limit these opportunities; see Figure 14–8.

Special efforts are needed in agricultural education, extension, and training to include a higher proportion of rural young women. Traditional home economics programs need to be broadened to include agricultural and income-earning skill training. Agricultural schools, colleges, and universities that prepare extension personnel need to make a greater effort to attract female students.

Elements Conducive to the Effectiveness of Agricultural Extension

Some of the elements conducive to the development and continuing effectiveness of extension services may be summarized as follows:

1. a national policy embodied in legislation establishing the educational role of the extension service and the relationship of extension education to other elements of agricultural and rural development.
2. a philosophy of extension education embodying the concept of human resources development as a major goal.
3. sources of technical information required for the solution of the problems of rural people.

---

*Adapted from the Food and Agriculture Organization of the United Nations, Rome, Italy.

Figure 14–8 • A child in Sierra Leone waters cocoa plants that will eventually be sold to farmers in her village. CARE assists farmers in forming village marketing associations and building market stores so they can get a better price for their cocoa and reduce transportation costs. *Source: Care; photo by Vera Viditz-Ward.*

4. sources of trained people adequate to supply the personnel needed in extension and other related services.
5. adequate **ancillary services** such as farm supply credit, transportation, marketing, and veterinary services, without which extension teaching is largely ineffective.
6. an environment conducive to continuing agricultural and rural development, including such things as incentives for production, political stability, and a land

tenure system which assures the producer his fair share of the benefits accruing from improvements in his farming practice.

Where should the extension service be placed in the organizational structure of the Ministry of Agriculture? Several factors should be considered in making this decision, and include the following:

1. Will the extension service be the educational arm of the Ministry of Agriculture?
2. Will the extension administrator be a part of the Ministry of Agriculture's policy-making staff?
3. Does the role of extension provide effective working relationships with the agricultural research agencies?
4. Will the operating procedures provide for good working relationships with all agencies of the Ministry of Agriculture involved with agricultural education and rural development?

Legislative Provisions Needed

A survey of legislation pertaining to extension work in 52 countries indicates that only 15 countries had laws which provided an extension budget specifically for their operations. About one fourth of the nations surveyed had no legal basis for extension work. Over one third were operating under the broader legislation establishing a Ministry of Agriculture. This study made several recommendations concerning provisions which are important to include in any extension laws, and are as follows:

1. Provide educational programs for the whole family.
2. Provide separation of educational and training work from enforcement and regulatory acts such as plant and animal quarantine, seed, feed, and fertilizer quality inspection and distribution.
3. Provide for freedom from political involvement.
4. Provide for close affiliation with the research programs of agricultural experiment stations and the training programs of agricultural institutes.
5. Provide for guarantee of financial support annually.
6. Provide for regional organizations geared to the needs of the various areas.
7. Provide for advisory and planning groups to adapt programs to the needs of the people and to the existing conditions in the area.
8. Provide for local and regional participation in extension programs.

The law creating and supporting extension must be compatible with the legal structure and social organization of the country. The law establishing extension does not guarantee an effective extension service—it only provides a legal foundation upon which to build the service. How the law is implemented, the status of the extension worker, and the financial support provided are some key ingredients for developing an effective extension service.

Resources Needed

There are several kinds of resources needed in order to establish and maintain a dynamic agricultural extension service. Essentially, these include the following:

- **human resources**—trained and dedicated personnel.
- **physical resources**—supplies, equipment, transportation, and meeting places.
- **scientific knowledge**—research results that can be adapted to agricultural and rural development.
- **financial resources**—to hire, train, and keep competent employees.
- **administrative and political support.**

Most people agree that extension work in agriculture can contribute much to the economic and social development of a nation. But, reaching the millions of rural families with this service is a major problem. There are two ways this service can be improved: directing more economic resources to the educational organization; and increasing the efficiency of the present system. Both approaches will need to be followed to some extent in most nations. As the extension service is expanded and improved, governmental revenues will undoubtedly increase, thus allowing more resources to be allocated for extension purposes.

Teaching Methods Adaptable For Education Through Extension

Considerable research regarding teaching methodology in extension has been conducted in developed nations. A minimal, yet growing, amount of data is accumulating from developing countries. These findings have established the comparative usefulness of certain methods and combinations of methods. Editors of CERES (July–August 1981) relate the following experience of INADES (African Institute for Economic and Social Development):

> From the beginning, three things were apparent. First of all, there were plenty of agricultural research institutes in Africa, but all had the same peculiarity of keeping the results of their work to themselves, since there was no organized way of spreading information. Secondly, there was no link between the agriculturist and the small farmer; there was need for an intermediary body of extension workers. Lastly, although many well-

intentioned people were trying desperately to help the rural world, their efforts were scattered. There was no idea of follow-up. Most extension workers were foreigners who returned home after two or three years without leaving permanent trace behind them.

From these observations, INADES defined certain lines of action and, in particular, two Golden Rules:

1. Never preach to the farmer about anything that he cannot immediately put into use. (For example, there is no point in talking to him about mechanized agriculture if neither his personal means nor the organization of credit facilities allow him to think of it, or if there are no spare parts or repair and maintenance services immediately available.)
2. Get through to the farmer in the language he knows, with words he understands. This means that, from the start, the involvement of extension service workers who speak the local language and may be recognized as "belonging."

Keeping in mind these two important rules, extension workers may choose a teaching method from the following brief descriptions.

▶ **NOTE** The following information is largely presented on a "how-to-do" basis and is adapted mainly from federal and state Cooperative Extension manuals. While these materials were designed and produced by workers in the U.S., virtually all of the practices recommended will, to some degree, apply to extension programs in other nations.

Selecting the Proper Teaching Methods

As the educator gains experience, he/she will develop skills in using proper teaching methods. The teacher should be the one to select the learning experiences that are to be provided. The methods to be used will depend on several factors, including the following:

- the number of people you are trying to reach.
- the size of the local and supporting staff available.
- the availability of certain **communication media,** such as telephone, newspaper, radio, and television.

The characteristics of the audience to be reached, and the complexity of the subject matter or skill to be taught must also be considered when making a selection.

Individual Contacts

Farm and Home Visits. Each visit should have a purpose. Since the visit is a heavy consumer of the teacher's time, you should prepare to make the visit effective. Secure as much background information on the situation as possible. Bring supporting information with you. When visiting the farm or home, see something to compliment, ask questions, and then lead into the problem. Strive to leave with a feeling of satisfaction for both you and the person visited. Be a good listener and try to identify other problems that you may help to solve.

Office Calls. Each office caller should be greeted curteously and made comfortable. The caller's needs should be promptly taken care of in a friendly and businesslike manner. Since the visitor has made a special trip to see the extension agent, he or she will probably be highly receptive to learning, and full attention should be given to his or her problem. For an office call to be successful, the visitor should leave with the information sought or be assured that he or she will receive it as soon as possible. The office files should be kept up-to-date with the reference materials readily available.

Result Demonstrations. "A man will remember what he learns from an equal long after his master's words are forgotten" (Kipling). The success of farmer's cooperative demonstrations in the U.S. was largely due to the unusual personality of Seaman A. Knapp, the "father" of agricultural extension programs. Knapp's philosophy was "What a man hears, he may doubt; what he sees, he may possibly doubt; but what he does for himself, he cannot doubt."

A **result demonstration** is a method of teaching designed to show by example the practical application of an established fact. Under the direction of the agent, the individual carries out the demonstration. Result demonstrations cover a period of time. Some comparisons are made, and records are kept. To be most effective, they must be seen by large numbers of people.

Plans for the demonstration should be in writing. The plan should include the number of demonstrations, location of demonstrations, material needed, kind of records to be kept, number of check plots for comparisons, plans for supervising the demonstrations, plans for field days, and plans to disseminate and use the results. The demonstrator should have farm or home conditions that are similar to those experienced by the audience to whom the demonstration is directed. This is a valuable teaching method for introducing new practices.

Group Methods

Method Demonstrations. The purpose of method demonstrations is to present an improved practice in an interesting, convincing way so that people will appreciate its desirability and its practical application to their situation.

All persons interested in the demonstration should be invited to attend. Publicize the demonstration in a news article, using local names and background information describing the new practice and the way it is done.

The demonstrator must be skilled and experienced. It is most important that the demonstrator know his or her job and how to teach it. An outline of each step in the demonstration should be developed ahead of time.

Steps in conducting the demonstration are as follows:

1. explain the need for this practice to the individual, family, or community, if necessary.
2. present the demonstration using simple words. If new words are used, explain them. Show operations slowly, step by step, and repeat if possible. Emphasize key points and tell why they are important.
3. invite help or participation from the group.
4. ask for questions and discussions.

Leader Training Meetings. Training volunteer leaders to help large extension education programs is an effective way to help people accept new ideas and practices. The leaders are trained by the extension worker to do a specific job.

After attending training meetings, the leaders give method demonstrations and talks at meetings, assist with tours and achievement days, and serve as leaders in organized youth groups.

Lectures. Lectures are a good way to impart new information. Some tips for preparing effective lectures include the following:

1. visualize the intended audience and their needs.
2. prepare more background facts about your subject than you will have time to discuss.
3. arrange facts in a logical order and use familiar words.
4. use good enunciation and clear visual aids.
5. believe in your message and speak enthusiastically.
6. when using numbers, use round numbers and only a few of them.
7. concentrate on the main ideas and end with a short conclusion.

Conferences and Discussion Meetings. Meetings, conferences, and group programs are the oldest and most important teaching aids we have (Kelsey and Hearne, 1963, p. 405). Many teaching tools may be used at meetings, including printed material, the spoken word, method demonstrations, individual participation in the activities, and visual aids. Many meetings take the name of the meeting objective, such as planning meeting, training meeting, organizational meeting, or achievement day.

During the meeting, pivotal questions should be raised to increase interest and understanding. Meetings should be held in convenient locations for the intended audience; often this may be under a big shade tree in the countryside.

Many times, discussion leaders are used to stimulate ideas and exchange of information. The function of a discussion leader is to get all points of view in the open, to keep the discussion on the subject, and to try to arrive at group decisions.

Tours. Tours are used frequently to allow interested people to observe what happened at a result demonstration. The purpose of the tour should be publicized. Advance arrangements should be made with the tour host, including the time and date of tour, and key points for him or her to emphasize at the tour stop.

Schools. The term school as used in adult education means a place for instruction (Kelsey and Hearne, 1963, p. 164). Schools are designed to give the participants new knowledge, skills, and attitudes. They are usually related to one subject matter area and may last from 1 to 5 days. The schools provide an easy way to present new information for a special interest group. The teaching can be organized in an effective, logical way, and a school offers the opportunity to use many teaching methods.

Miscellaneous Meetings. There are other types of meetings that are held by the adult educator in the operation of his or her programs. These may include special *ad hoc* committee meetings, budget meetings, and meetings to inform specific people about the educational programs planned and completed.

Mass Media

Mass media makes possible the dissemination of information to a large group of people. The use of different mass media tools such as radio, television, circular letters, exhibits, and posters provides for helpful repetition. One of the most effective mass media techniques is the use of the publication.

Publications such as bulletins, leaflets, fact sheets, and circulars play an important role in providing information to clientele. If written at the educational level of the target audience, publications have many advantages over other mass media, including the following:

- people have confidence in the printed word.
- publications provide accurate and detailed information.
- they may be used as a substitute for personal letters.
- a well-written publication creates interest for additional information.
- the extension agent can use subject matter from a bulletin as a news article.
- printed materials can be filed for future reference.

An attractive display of bulletins should be available for inspection, and bulletins should be announced through the mass media.

Conclusion

Agricultural extension personnel who are involved in on-farm tests and demonstrations or surveys must continually improve their knowledge and skills. Farm testing and demonstrations provide a means of training new production specialists (as well as farmers) who need experience in practical farming, as well as knowledge of technology and its application. For more experienced personnel, a continuing program of farm experimentation allows them to keep up with new technological developments.

Government agricultural extension services must go beyond merely increasing the flow of technical information to the cultivator. Farmers often need assistance in obtaining the inputs necessary to take full advantage of the new technologies, including credit at reasonable rates or interest with reasonable procedures, and fairly priced, adequate amounts of fertilizers, pesticides, and herbicides when needed. Access to markets with stable crop prices is necessary to ensure favorable relationships between input costs and crop prices. These problems can often be resolved if they are fully understood and properly presented to the people. The extension agent should be sensitive to providing this type of assistance in response to farmer's suggestions, or on his own initiative. Effective training to improve skills, knowledge, and competencies can be provided to farmers/producers by extension personnel.

Discussion Questions

1. What is the rationale for advising governments in other nations to allocate resources to teaching and extension? Is it important to the people of the U.S. that a dynamic extension service be maintained? Why or why not?
2. Briefly describe the organization and function of the Cooperative Extension Service of the U.S.
3. Explain why the role of the extension field worker (extension agent) is so crucial to the maintenance and development of a viable agriculture sector.
4. Briefly discuss the organizational structure and policies that may enhance the productivity of an agricultural extension service.
5. What factors would influence selection of those teaching methods that would be best adapted to extension work with small farmers?
6. Explain the difference between method demonstration and result demonstration in terms of usage and situation adaptability.
7. What are some limiting factors confronting some nations' extension programs?
8. Briefly describe the role of women in agricultural production.
9. Why do rural youth of all nations require agricultural education, extension, and training programs?
10. What are some major elements conducive to the effectiveness of agricultural extension programs?
11. List and explain each of the kinds of resources needed in order to establish and maintain a dynamic agricultural extension service.
12. Why is it important that an extension agent develop skills in using proper teaching methods?

CHAPTER 15

Agricultural Research—A Major Force in Development

Source: United Nations Development Program; Photo by Ruth Gassey.

Objectives

After reading this chapter, you should be able to:

- discuss the nature and extent of agricultural research as it has developed and now functions in the U.S.
- state the objectives of the U.S. Nationwide Cooperative USDA State Agricultural Research Programs.
- discuss the needs for research in international agriculture development, specifically in the area of food production technology and resource management.
- list and locate each of the International Research Centers and describe the primary function of each.

Terms to Know

Mission-oriented
cooperative USDA/state research program
Consultative Group on International Agricultural Research (CGIAR)
CIMMYT
IRRI
International Agricultural Research Institutes
IFPRI
basic research
supporting research
strategic research
tactical research
farm-level operational research

Introduction

As man finds it necessary to modify and maintain environmental factors suitable for the production of foodstuffs in order to meet the increasing demands of the world's population, attention to scientific investigation becomes almost mandatory. Not only is it necessary to discover effective techniques of culture and profitable practices in environment modification and maintenance, it is also vital to increase the innate productive capabilities of agricultural species, both plant and animal.

Higher yielding varieties of food crops and more productive breeds of livestock are often the product of painstaking applications of the science of genetics. The discovery of a dwarfing gene in wheat and rice brought about higher-yielding varieties of these crops. These research breakthroughs, in association with higher-level production management practices, have largely brought about the "green revolution."

In a similar manner, agricultural scientists have been able to manipulate genes in livestock to an extent that strains are developed that can metabolize feeds more efficiently and produce greater quantities of milk and meat and more desirable cuts.

Other inheritable traits include a resistance to parasites and disease. More efficient technology usage, better design of equipment, and alternative energy sources are also of great importance. Careful examination of the feasibility and efficiency of alternate farming systems is essential as emphasis shifts in the direction of strengthening the indigenous small farmer.

Agricultural Experimentation in the U.S.

The U.S. ranks high in publicly supported research directed toward producing more food, securing needed resources for agriculture, and maintaining an adequate delivery system. Strengths of the U.S. include large size, diversity, and experience in practical problem solving.

The agricultural research system involves half a dozen federal agencies and more than 50 state research organizations. The publicly supported agricultural research programs are supplemented by a considerable number of private industrial research organizations. If current levels of support are continued in the future, and necessary resources continue to be allocated, the U.S. agricultural research system will be able to respond effectively to the new challenges of increased food demand both at home and abroad.

A salient factor in the success of the U.S. agricultural research system is that it operates as an integral part of the "Land Grant" system in which instruction, research, and extension education constitute combined efforts organized and operated within the agricultural university and college structure. The U.S. Department of Agriculture must also be recognized as a positive influence on the successful performance of the research sector within the agricultural complex.

Public agricultural research is conducted mostly within the U.S. Department of Agriculture and at state Land Grant Universities, both of which receive federal funding.

Agricultural research is conducted primarily at fifty-five state agricultural experiment stations where research is interwoven with the training of new agricultural scientists.

U.S. Research System—Mission Oriented

It would be good to examine the basis that the U.S. research system is mission-oriented, as well as to review the several objectives that have been established for that system. In 1976 and 1980, landmark reports were issued by consultative and professional groups which specified the function of and expectations for the Nationwide Cooperative USDA-State Research program.

The public agricultural research system in the U.S. is **mission-oriented,** which means that it must be responsive to society's needs for an adequate food supply and for effective use and conservation of resources. But agricultural research also faces the demands and dilemmas of the scientific process.

In the U.S. public agricultural research systems, both the scientific community and society's policymakers are involved. Sometimes it is not entirely clear why policymakers do not see the importance of research in certain areas. Policymakers, on the other hand, often ask why scientists do not see the importance of clear and relevant research reports without undue qualifications.

Much of the research community's efforts are directed at specific goals and targets. However, another role of the scientist is to build a store of basic knowledge and methods that will provide answers to problems of the future. The chief role of the policymakers in either case is to evaluate the research plans and projects in light of society's most pressing needs.

Objectives of the USDA/State Research Program

The nationwide **cooperative USDA–state research program** has several objectives, which include the following:

- to solve local, regional, and national problems affecting agriculture, forestry, and other renewable natural resources. This includes work toward an adequate and safe food supply for all consumers; protection of environmental quality; and, quality of rural life.
- to provide a continuing flow of new scientific knowledge essential to the solution of future problems.
- to provide scientific competence for teaching, including graduate student research, in order to train future generations of scientists.
- to provide scientific expertise to local, state, and federal agencies; private organizations and individuals; and, programs of overseas agricultural development.

The United States private industry in the U.S. spends large amounts on agricultural research and development, primarily to promote technological development. Part of these funds are in support of public research.

Agricultural Research in Other Nations

Agricultural research efforts and programs vary greatly throughout the world, particularly among the developing nations. Every nation needs some research capacity. Especially important is the capacity to select from available experimental lines of major food crops and livestock breeds in order to best adapt farming methods to the environments in which they are to be used. However, lack of funding continues to hamper research programs in most of the developing countries.

The research efforts of virtually all of the developing countries suffer from a common handicap. National investment in support of agricultural research, education, and production is low by percentage of GNP, per capita per year, or value of agricultural production. Thus, most developing countries are unable to cope successfully with their difficult agricultural problems without external assistance.

Typically, developing countries lack trained manpower; strong, problem solving, and development-oriented research institutions; and sound national agricultural research policies and procedures.

The will to implement such policies, as shown by inadequate salaries and research support, also frequently fluctuates over time with changes in government.

Consultative Group on International Agricultural Research (CGIAR)

The **Consultative Group on International Agricultural Research (CGIAR)** is an informal association of over 40 countries, international and regional organizations, and private foundations established in 1971 to support a system of agricultural research around the world. At present there are 13 international agricultural research centers in the CGIAR system, most of them located in developing countries. The centers are independent and autonomous. Each has its own governing board, charter, and research agenda.

The World Bank, the United Nations Food and Agriculture Organization (FAO), and the United Nations Development Program (UNDP) are cosponsors of CGIAR.

Over 600 senior scientists representing 60 different nationalities conduct research either at one of the 13 CGIAR centers or in 36 other developing countries. The approach is problem-oriented and multi-disciplinary. The centers' scientists combine their experience and expertise to find the causes of low productivity and determine potential cures, taking into account the local conditions under which a particular food commodity has to be produced.

Crop improvement, by which desirable characteristics are developed within a plant, accounts for 70 to 85 percent of research expenditure in the CGIAR system. Other research areas include agronomy, economics, engineering, farming systems, animal nutrition, and animal diseases.

The most noted success for the CGIAR-sponsored research centers was the development of high-yielding varieties of wheat and rice that were the backbone of the green revolution in Asia. Those improved varieties were the product of research at **CIMMYT** (for wheat in Mexico) and **IRRI** (for rice in the Philippines). High-yielding varieties of rice and wheat are now grown on an estimated 115 million hectares throughout the world, producing 40 million more tons of food grain per year than could be provided by traditional varieties with the same inputs. This increase is enough to feed half a billion people.

Improved maize and field beans are also having an impact on food supplies. Over 200 improved varieties of maize, combining increased yield potential with resistance to pests and diseases, are now being grown on more than six million hectares in the developing world, and over 100 improved bean lines have been released to farmers by national research systems that are partners of the CGIAR centers.

Over 250 improved varieties of sorghum, potato, cassava, chickpea, cowpea, pasture species, pearl millet, pigeonpea, and durum wheat have been released by national programs in regions where these commodities are in demand.

In addition to crop improvement, other examples include broad-bed-and-furrow cultivation, integrated pest management, genetic resource missions, and the training of over 20,000 scientists.

International Agricultural Research Institutes (The 13 CGIAR Centers)

One component of the international agricultural research system is formalized in a single organization, the **International Agricultural Research Institutes.** The basic idea underlying the institute system is to have a core group of scientists, including breeders, geneticists, agronomists, engineers, entomologists, and others working in a central location with good facilities and adequate resources. Findings of the institutes are then tested at many locations around the world. The International Agricultural Institute system began when the International Rice Research Institute (IRRI) was established in the Philippines in 1960. Now 13 institutes are operational. The work of the institutes not only fosters and facilitates international cooperation, but greatly enhances release of information and selection of practices best suited for specific environments. As a result, the institute system is providing a very important means for hastening the development of new and improved agricultural technologies. Figure 15–1 depicts the location of each of the institutes within the International Agricultural Research Network.

Figure 15–2 provides a listing of each of the International Agricultural Research Institutes, their location, emphasis of research initiatives, geographical coverage, and date of initiation.

252 Chapter 15

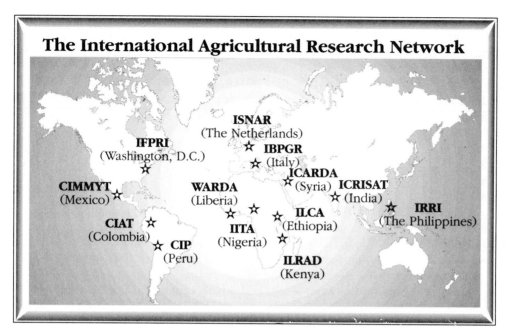

Figure 15–1 • The International Agricultural Research Network.

The basic purpose of the International Agricultural Research Institutes is to help nations develop their capabilities to increase agricultural production. Toward this end, institute staff members work cooperatively with those of national institutions on the more difficult problems of regional or international importance, train staff members for national institutions, and facilitate international cooperation on matters of importance. Experimental strains and practices developed at the institutes are tested at many locations in the world along with materials developed in cooperating national programs. This permits selection of varieties best suited for specific environments. The international institutes make germ plasm freely available to cooperating countries, provide scientists to work cooperatively in national programs, hold meetings to review programs and exchange information, and train national scientists. Similar work is being carried out in animal production.

Even though the international institutes perform a very important role, they are no substitute for strong national programs if the production of food and fiber in the world is to be increased. For best results, national programs need to be closely related to institute programs to the extent that scientists in each nation can have access to an institute for periods of training and for participation in short courses to update techniques and planning.

An especially important center within the CGIAR system, in terms of recommending policy for enhanced agricultural development, is the International Food Policy Research

The International Agricultural Research Institutes

Centre	Location	Research	Coverage	Date
IRRI (International Rice Research Institute)	Los Banos, Philippines	Rice under irrigation, multiple cropping systems and upland rice	Worldwide, special emphasis on Asia	1960
CIMMYT (International Centre for the Improvement of Maize and Wheat)	El Batan, Mexico	Wheat, triticale, barley and maize	Worldwide	1964
CIAT (International Center for Tropical Agriculture)	Cali, Colombia	Cassava, field beans, rice and tropical pasture	Worldwide in low land tropics, special emphasis on Latin America	1965
IITA (International Institute of Tropical Agriculture)	Ibadan, Nigeria	Farming systems, cereals, cowpeas, soybeans, lima beans, pigeon peas, cassava, sweet potatoes and yams	Worldwide in low land tropics, special emphasis on Africa	1965
CIP (International Potato Centre)	Lima, Peru	Potatoes (for both the tropics and the temperate regions)	Worldwide, including linkages with developed countries	1972
ICRISAT (International Crops Research Institute for the Semi-Arid Tropical)	Hydersbed, India	Sorghum, pearl millet, pigeon peas, chick peas, farming systems and ground nuts	Worldwide, special emphasis on dry and semi-arid tropics, non-irrigated farming	1972
ILRAD (International Laboratory for Research on Animal Diseases)	Nairobi, Kenya	Trypanosomiasis, theileriosis (mainly east coast fever)	Africa	1974
ILCA (International Livestock Centre for Africa)	Addis Ababa, Ethiopia	Livestock production systems	Major ecological regions in tropical zones of Africa	1974
IBPGR (International Board for Plant Genetic Resources)	FAO, Rome, Italy	Conservation of plant genetic resources	Worldwide	1973
WARDA (West African Rice Development Association)	Monrovia, Liberia	Regional cooperative effort in adaptive rice research among West African countries	West Africa	1971
ICARDA (International Centre for Agricultural Research in Dry Areas)	Aleppo, Syria	Barley, lentils, broad beans and farming systems including animal husbandry	Emphasis on dry land farming and arid areas in Near East and North Africa	1976
IFPRI (International Food Policy Research Institute)	Washington, USA	International food policy and food distribution	Worldwide	1975
ISNAR (International Service for National Agriculture Research	The Hague, The Netherlands	International service organization	Worldwide	1979

Figure 15–2 • The International Agricultural Research Institutes.

Institute (**IFPRI**). Membership in the Institute includes noted economists from many countries who focus on strategies and plans to meet world food needs. Research includes all aspects of policy analysis. The word "policy" in its name should not be overlooked, for IFPRI is responsible for recommending to countries those policy options that will ensure that people everywhere have adequate food and nutrition. Almost all of the staff economists come from the developing countries; when they return to their homes, they are expected to provide a link between IFPRI and the important centers of research in these countries.

The Nature and Scope of Agricultural Experimentation in Developing Countries

Research efforts in any country may be classified as follows:

Basic research is research undertaken to develop knowledge for its own sake.

Supporting research is fundamental but purposeful investigation for probable usefulness of findings only partially known.

Strategic research is aimed at solving those problems which affect several regions of the world, or indentifying new varieties or practices for use in those regions.

Tactical research is undertaken for the specific purpose of identifying improved components of farming systems that change-agents can use to meet needs of farmers in a particular activity.

Farm-level operational research involves the identification, through experimentation on farms, of the specific combinations of crop and animal production practices that will provide maximum productivity on those farms.

Each of these five types of research is needed to some degree in all nations. However, there is a tendency in many countries to overlook or even neglect tactical and farm-level operational-type efforts. This may be due to the bias of researchers toward more sophisticated activities which, they believe, may bring more prestige to the investigator and the nation in which he operates. Whatever the reasons, maximum benefit from research activities worldwide depends upon closer coordination among research agencies, with an emphasis on problems that can be identified as occurring at the operational farm level; see Figure 15–3.

Conclusion

There is sufficient evidence to conclude that farmers, even those who are uneducated and have small holdings, will readily make use of research findings, particularly those coming from tactical and farm-level operational research, provided that the following four conditions are met:

1. There must be available to them more productive and more profitable farming systems, and the systems must be complete.

Figure 15–3 • A substantial increase in dairy production is needed in India to provide more nutritious milk products for its rapidly increasing population. The Food and Agriculture Organization is assisting India to establish several centers for advanced dairy studies at the National Dairy Research Institute in Karnal. The newest center on dairy animal biotechnology research provides post-graduate and training programs in molecular and genetic engineering, improved feed resources, and hybridization of cattle, buffalo, and goats for increased milk production. *Source: Food and Agriculture Organization of the United Nations; photo by I. de Borbegyi.*

2. The necessary inputs—fertilizers, seed, pesticides, and credit—must be available to the farmer when and where he needs them, and at a reasonable price.
3. The farmer must be shown how to effectively utilize the new technological system.
4. The farmer must know before he invests in new plantings or other operations that there will be at harvest a market for his product at a price on which he can depend. A market requires roads, transport, effective demand for products, and favorable prices.

Discussion Questions

1. Explain each of the following types of research: basic; supporting; tactical; strategic; and, farm-level operational.
2. How do agricultural research efforts function in the U.S.?
3. Discuss the mission and objectives of the U.S. publicly supported research system.
4. Explain the organization, function, and give the location of the International Agricultural Institutes, and indicate what role they play in furthering research efforts throughout the world.
5. List the areas of agricultural research judged to be of highest priority in efforts to insure adequate world food supplies.
6. Why is the need so critical to promote and maintain research in crop and animal production, and soils management in tropical and sub-tropical areas?

CHAPTER 16

Design and Implementation of Agricultural and Rural Development

Source: DeKalb Plant Genetics

Objectives

After reading this chapter, you should be able to:

- define agricultural, rural, community resource, and community/village development.
- state the relationship between agricultural development and community development.
- discuss the advantages of small-scale, on-farm experimentation.
- discuss agricultural cooperatives as a strategy to foster agricultural and community development, and explain what precautions must be taken if cooperatives are to be of maximum benefit.

Terms to Know

community resource development
participative program planning
small-scale experimentation
commodity-oriented production programs
defined-region campaigns
synchronized government services
cooperatives

Introduction

Around the world many nations are entering an era of agricultural and rural development that is characterized by their government's heightened awareness of the need to accelerate agricultural and rural development. Knowledge, financial resources, and management capabilities are being mobilized to achieve breakthroughs in crop and animal productivity, as well as to advance human welfare.

The success of these development activities will be dependent on such factors as public investment in roads, rural electrification, communications systems, and other services necessary for an advancing agriculture. The design and implementation of agricultural and rural development is dependent upon education, extension programs, and research.

Agricultural and Rural Development Defined

Agricultural development expands by a process of growth, and makes available or usable its resources. Development involves the attainment of goals and objectives. At the higher levels of operation (the international and national scenes) agricultural development is largely viewed in terms of getting resources, both human and material, in a pattern that will enhance the acquisition of an adequate supply of food for each nation.

Major organizational issues in rural development include determining relative priorities in resource allocation for the different sectors of the economy, and for social services, sequencing these priorities, and effecting policy and institutional changes for maximizing resource effectiveness. Many people with experience in rural development believe only when there is a strong government commitment will enhanced productivity come about.

Development assistance in agriculture often includes a considerable portion of the total assistance that one nation may grant to another. Of the total monies expended worldwide annually by the U.S. Agency for International Development, approximately 27 percent is allocated to agriculture. For example, AID-proposed expenditures for Latin America during the 1980s included the following:

- agriculture, rural development, nutrition: $127 Million
- population planning: $13 Million
- health: $23 Million
- education and human resource development: $40 Million
- selected development activities: $71 Million

Emphasis upon and sufficient resource allocations to agriculture, rural development, and nutrition continues to be on agricultural research, extension, and supporting activities that will almost invariably result in greater food production, higher incomes, and increased employment for small farmers and other people in rural areas.

Another example of an agricultural development project includes a new watershed management project in Peru that will introduce irrigation systems, soil conservation, forestation, and crop improvement programs in the sierra and high jungle areas. This is particularly important since only 3 percent of Peru's land base is considered suitable for intensive agriculture. Other new projects include agricultural education in Jamaica and agricultural planning in Panama. Ongoing projects will continue to stress conservation of natural resources, forestry, fish ponds, and promotion of agriindustrial activities.

It should be emphasized that 46 percent of monies allocated by nongovernmental U.S. institutions are expended for agricultural development. The largest single unit allocation in the total budget of the United Nations Organization is for agricultural development.

The term rural development is often used in association with agricultural development. However, the terms are not always synonymous since, in rural development, certain other areas of concern are generally included in addition to agricultural production, agricultural financing, and agricultural planning and management. Rural development in the U.S. covers all types of activities in nonurban areas, including cities of 50,000 or less, and suburban or rural areas surrounding cities with larger populations.

Rural development includes loans for community facilities and business ventures, some of which are made by private lenders. Rural development can include a new office, plant, or laboratory providing local jobs for rural people, and can include loans to enable people to fix up their homes.

Rural development, or the synonymous term, community resource development, is a process whereby those in the community arrive at group decisions and take actions to enhance the social and economic well-being of the community. This process is concerned with what happens to people as a result of group decision-making and the implementation of programs concerning the community's social, economic, and institutional components. It analyzes the forces of change, including movements in population, shifts in types of employment, the role of government, and the emergence of new social and economic organizations.

Community resource development is the process of identifying the leadership in a community, making citizens aware of the opportunities and alternatives available, and providing requested assistance. This process is inclusive of all levels of government—local, state, and federal—and is dependent upon available resources and the involvement of the local people.

Program Planning for Agricultural Development

There is no blueprint for rapid agricultural development which is applicable to all countries. But for virtually every region or problem, certain positive actions can be initiated.

A key ingredient for success is **participative program planning.** An effective planning process should consist of the five following basic stages:

1. gathering information.
2. establishing goals and objectives.
3. developing a work plan.
4. preparing the budget.
5. monitoring.

In practice, these are not distinct successive stages. Planning inevitably moves in a series of loops and digressions, with two or three stages often proceeding simultaneously. For example, monitoring may suggest a new work plan or the need to gather more information, while budgetary limitations may necessitate a lowering of objectives or the adoption of a less expensive work plan.

An important point about planning is that local clients should be involved from the initial stages throughout the project. It is also essential that the producers be involved in certain crucial determinations as follows:

1. Does the technology associated with practices for proposed adoption fit local farming and community living patterns for agriculture?
2. Do the technological changes deal with those factors that may limit production?
3. Does the improved practice promise benefit to the producer, and does the technology utilize those resources the producers already have?

Farmers, regardless of size of landholding, generally will increase their productivity provided four requisites are met, as follows:

1. An improved farming system. A combination of materials and practices that are more productive and profitable, with an acceptably low level of risk, than the one currently used must be available to the farmer.
2. Instruction of farmers. The farmer must be shown how to put the practices into use, and should understand why those practices are better.
3. Supply of inputs. The inputs required and credit to finance their purchase must be available to the farmer where it is needed and at a reasonable cost.
3. Availability of markets. The farmer must have access to a nearby market that can absorb increased supplies without excessive price drops.

Other Features of Successful Agricultural Development Programs

Of considerable importance in fostering successful agricultural development is the practice of **small-scale experimentation.** This method is generally applied to the practices of each farmer, although the same principle may well be related to larger community, area, national, or even cooperative international research efforts. Even so, the essential effort is to be made by the individual producer as he develops his own farming program. Bunch (1982, p. 139) enumerated the advantages of the small-scale experiment.

The first advantage for the small farmer is that it reduces risk and protects against major economic failure. Innovations can fail for many reasons. They may not have been tested sufficiently by the program. Those that have been well tested can fail because of differences in weather, topography, microclimate, or soils between the time and place the innovation was tested and when it was put into practice. If the farmer starts by trying the innovation with a limited quantity of land or with two or three small animals, the methods and the probable results will be known before an entire year's income is risked. If there is a loss, it may hurt, but it will not affect economic well-being for months and years to come.

A second advantage of small-scale experimentation is that a farmer can learn much more than by experimenting with the entire crop. If a farmer makes a change in the entire crop or all of the animals, only one change or one combination of changes can be tried each year. For example, one farmer experimented with three different vegetables, five varieties of pasture grass, three or four soil conservation methods, and various plant populations for corn all at the same time.

Thirdly, a farmer who makes a change in an entire crop or with all of the animals has no way of comparing the results of the new production system with those of the previous one. If the harvest or the animals improve, the farmer may never be sure whether the improvement is due to the innovation itself or to fortuitous circumstances, such as good weather or less disease. On the other hand, even a very good innovation can fail when conditions turn unfavorable. If, however, a farmer tries out the innovation on a small scale, the rest of the farm is a natural control plot.

A number of scholars and researchers who specialize in agricultural development likewise stress the importance of small-scale experimentation.

Another necessary strategy for successful agricultural and community development is realistic goal setting with objectives carefully worked out through joint participation by teacher/promoters and each farmer involved. Some countries have been surprisingly successful in applying a simplified adaptation of the "Management by Objectives" system to programs for increasing agricultural production, as well as achievement in the areas of sanitation, health care, and other areas of community living.

Although the key to successful agricultural development is the involvement of the producers in planning, implementation, and management, oftentimes the nature and scope of governmental involvement is likewise a crucial factor. Wortman and Cummings (1978, p. 241) pointed to three well proven and complementary approaches available to governments as follows:

- **commodity-oriented production programs** designed to achieve established goals for domestic consumption or export.
- **defined-region campaigns** to increase productivity and incomes of as many people as possible, using whatever combinations of commodities, techniques, and services are feasible.
- **synchronized government services** to speed progress.

For greatest success, the strategy should combine all three approaches. The relative emphasis on each approach, and the ways in which each is implemented, will be dictated by the goals established.

The commodities to be emphasized and the production goals to be met are decisions of major importance. Governmental bias toward production of crops for export rather than food crops for home consumption is always a hazard that must be recognized. A recommended procedure to deal with these conflicting interests involves government bringing together a carefully selected and widely representative task force to establish priorities and to design a program. As the program advances to the implementation stage, the matter of securing the approval and cooperation of various groups and agencies is essential.

Cooperation among organizations and individuals should be a central feature of commodity programs. Field stations, production centers, colleges or schools of agriculture, and farms should participate in nationwide field trials and demonstrations. Producer groups can help arrange farmer participation in tests and demonstrations, promote interest, and identify logistical problems. Faculty members of educational institutions can undertake part-time research or promotion if funds to supplement salaries, pay assistants, and cover other expenses are available. The media should be used to get timely information to rural people and to keep the public informed of activities.

Local Cooperatives as a Function of Agricultural and Community Development

Local units in the form of **cooperatives** have often given farmers more solidarity. Cooperatives may provide marketing services, facilitate the securing of necessary credit, promote shared use of implements, provide for more equitable distribution of resources such as irrigation water, and promote the acquisition of goods and services with the advantages of pooled ordering and bulk deliveries.

Technoserve, a PVO (Private Voluntary Organization), reported on one traditional type of cooperative and two unique ones. The cooperatives are located in Kenya, Honduras, and Nicaragua.

For Kenya, the Harambee Savings and Credit Society, the largest savings and credit society in Africa, approached Technoserve for help in devising a new accounting system, establishing financial and management controls, evaluating staff, organizing training programs, and completing audits. As a result of improvements made, it was possible to end a moratorium which had been placed on new loans.

For Honduras, the Empresa Corocera Hondurena (ECOH) in Tela, Department of Atlantida, is exploring new processing and marketing systems for products derived from the corozo palm tree, which grows wild in the area. A new machine has been developed for cracking the corozo nut. Technoserve hopes to continue its efforts to find the right combination of human, organizational, marketing, and technical factors to insure the economic viability of the enterprise.

For Nicaragua, the Somoto Henequen Project, in the Department of Madriz, will expand its facilities for processing the henequen plant. Systems have been developed for the decortication of the leaf and extraction and sale of the fiber, which is used in the manufacture of rope. A machine has been installed to salvage waste fibers that can be used to manufacture mattress pads. Seedlings are offered to farmers to encourage increased production of this plant.

The continual function of cooperatives has not always met with approval from producers in the developing countries, particularly from small farmers. This is especially true of large governmental-sponsored cooperatives. The practice of granting subsidies to farmers, both in the form of production inputs such as fertilizers, and in concessional interest rates, too often becomes a serious drain on the economy. In some countries, powerful economic forces such as the large landholders tend to profit far more from governmental aid to cooperatives than do the needier small farmers. Also, a major problem for many cooperative units is the lack of trained managers to function in a leadership capacity. The great need for educated leadership among agriculturists is particularly evident when consideration is given to increasing potential for success of producer cooperatives in developing countries.

In more developed countries like Korea, agricultural cooperatives have been highly successful.

Agricultural Cooperatives in Korea

Thirty years have passed since the Korean agricultural cooperatives were organized in 1961 to enhance the economic and social status of farmers through voluntary organization of farmers. Despite the many difficulties that rural communities faced, the Korean agricultural cooperatives have contributed greatly to the development of those rural communities.

The agricultural cooperatives in Korea are a self-help organization voluntarily established and jointly capitalized by farmers to increase agricultural productivity, to enhance the economic and social status of farmers, and to construct high-income welfare rural communities.

With a view to attaining these objectives, the agricultural cooperatives have been conducting diverse business activities, including guidance for the increase of farm income and improvement of living; marketing of farm products; supply of farm inputs and consumer goods; banking and credit to mobilize and extend various agricultural funds; and, cooperative insurance for stabilization of livelihood and welfare of member farmers.

Conclusion*

Despite major progress in the development of agricultural technology, much remains to be done. Large gaps exist between actual and potential agricultural production, even among nations who have adopted new technology. Gaps also exist between experimen-

---

*Adapted from the National Association of State Universities and Land Grant Colleges.

tal and actual yields. Extension of present knowledge must become more widespread. A scientific research base is the key to the long-term success of the agricultural sector. Agricultural production is a dynamic process that requires continuing flow of new information concerning sophisticated issues. The problems of each nation will have to be diagnosed and solved. Although technical assistance must come first from scientists, local scientific capability must ultimately be developed. A long-term commitment is essential to ensure the creation of an effective and productive educational, extension, and research program. More specifically, research should be tailored to the requirements of each country or region, the private sector should be moved in more socially productive directions, and researchers should develop production practices and machines geared to the farm sizes of each country.

Discussion Questions

1. Define agricultural development, rural development, and community resource development.
2. How does the practice of teaching farmers small-scale experimentation prove effective in terms of agricultural development both in the U.S. and abroad? What are some of the advantages of small-scale experimentation?
3. What are the advantages of commodity-oriented production programs in stimulating agricultural development?
4. Why is participative program planning an effective ingredient for success in any type of development?
5. What are some of the advantages to farmers as a result of them forming their own cooperatives?

CHAPTER 17

Transforming Knowledge, Skills, and Commitment—Summing Up and Moving Forward

Objectives

After reading this chapter, you should be able to:

- categorize some causes and solutions with regard to worldwide agricultural production and inadequate nutrition problems.
- describe the current situation with regard to world agricultural production.
- discuss the association between poverty and hunger, and the major influences affecting each.
- discuss what is meant by the food-poverty-population-illiteracy problem.
- explain why it is important that all people receive instruction in agriculture.

Terms to Know

economo-realist
politico-parochialist
machino-technologist
socio/religio-humanist

behavioro-educationist
absolute poverty
relative poverty

civil conflict
world trade & economic
   development

Introduction

In concluding this text, an attempt will be made to:
1. summarize the information presented previously.
2. establish some pertinent facts about the current situation with regard to worldwide agricultural production and nutritional problems.
3. categorize some alleged causes and proffered solutions.
4. review certain controversies regarding the various approaches toward hunger alleviation.
5. establish a brief set of recommendations regarding policies and actions that are needed by all nations if substantial progress is to be made in increasing worldwide agricultural production, processing, and distribution.

The Persistence of Nutritional Problems

Because of a persistent lack of nutritious food, millions of people die each year or continue to live a meager existence. This situation is an insult to all humans because it denies the dignity and personal worth of each person born into this world. At least 150 million individuals are severely undernourished or malnourished in the sense that they do not have the energy to function normally. Hundreds of millions more would consume more and better food if they could. Yet, enough grain is produced throughout the world almost every year to provide every man, woman, and child on earth with the required calories for a healthy life. (Of course, even if all this grain were available, protein deficiencies might still be a problem.)

Luther Tweeten, an agricultural economist at Ohio State University, summarized the current situation with what he presents as eight propositions describing the world situation, and are as follows:

1. World food production per capita is greater than at any time in history, and more people have food to eat.
2. More people consume inadequate amounts of food today than at any time in history.
3. The source of these seemingly inconsistent propositions and the crux of the global malnutrition problem is uneven distribution of resources, rather than low volume of food production and consumption.
4. The growing nutrition deficit measured by the shortfall of food intake below requirements in developing countries will have to be closed largely by improvements in food production and distribution in the countries experiencing such deficits.
5. If distribution problems can be resolved, the world's resources and technology are adequate to provide satisfactory diets for all inhabitants of the world for at least the remainder of the twentieth century.

6. The major impediment to eradicating world malnutrition is man, not nature.
7. The world undernutrition problem reflects a broader income problem, and policies to deal with the food problem must also address the poverty problem.
8. The long-term solution to the world food problem lies in reduced population growth.

Positions Taken by Policy and Action Advocates of Agricultural Production

One way of analyzing and eventually synthesizing the various recommendations made by scholars and researchers in the area of agricultural production is by categorizing the individual scholar/researchers according to their predominant philosophies, beliefs, and viewpoints. These individuals have been classified based on the distinct and prevailing orientation of each toward agricultural production. These five categories are described in the following sections.

The economo-realist. The **economo-realist** is oriented in thinking toward a set of axioms or economic truths that, when applied to resources, work, supply, demand, trade, finances, energy efficiency, and other factors, will bring about predictable outcomes. Individuals in this category believe that programs of assistance to other nations should be based primarily upon these axioms.

The politico-parochialist. The **politico-parochialist** tends to see the world in relation to his or her own country and to judge as successful those efforts toward agricultural production that, while helping citizens of other countries to enjoy better living, will also enhance the political and economic security of the politico-parochialist's own country.

The machino-technologist. The **machino-technologist** possesses, foremost, a prevailing faith that technology holds the answer to providing plentiful nutritious food and to enhancing the quality of life for all people.

The socio/religio-humanist. The **socio/religio-humanist** is strongly rooted in the concept of the supreme value of each human life, regardless of race, creed, or origin. While motivation may differ between religious and nonreligious humanists, the end goal can be viewed as essentially the same.

The behavioro-educationist. The **behavioro-educationist** views agricultural production fundamentally as a matter of the developed countries promoting and enhancing the acquisition of knowledge and skills by peoples of the developing countries.

Obviously, no individual adheres to one of these philosophies exclusively, but one can identify a prevailing belief of viewpoint congruent with one, or seldom more than two, of these categories.

Causative Factors Associated with Inadequate Agricultural Production

Widespread agreement has been reached by scholars regarding causative factors associated with inadequate agricultural production.

Widespread Poverty

Financially impoverished farmers today are unable to acquire the necessary resources to produce, process, market, and distribute food.

Further investigation between poverty and hunger is supportive of policies to reduce conditions of underemployment and a stagnant economy. A nation's poor people cannot afford to purchase food. They have little (if any) means of earning an income. In brief, both the producer and the consumer must strive to enhance their quality of life.

According to Minear (1975) the main distinction among the poor, however, is not that between rural and urban people. It is between absolute and relative poverty. **Absolute poverty** is a condition of life so degraded by disease, illiteracy, malnutrition, and squalor as to deny its victims basic human necessities. The absolute poor are not merely a tiny minority of unfortunates nor a miscellaneous collection of the losers in life. On the contrary, they constitute roughly 40 percent of individuals living in developing nations. **Relative poverty,** on the other hand, means simply that some countries are less affluent than other countries, or that some citizens of a given country have less personal abundance than their neighbors.

The chronically hungry are mainly the absolute poor, caught not only without food, but without basic health, education, and employment opportunities. They are, says the United Nations, the victims of "the extreme inequality in the food distribution between socioeconomic groups" within a given country, which, as we shall see, mirrors a similar inequality between rich and poor countries at the international level.

Insecurity of Food Supplies

Statements claiming that enough grain is produced annually worldwide to provide adequate calories for each and every inhabitant of the planet must have a very hollow ring to the 150 to 200 million people who are severely undernourished and malnourished.

Food security means assured access to food. Problems of food security can occur on several levels, and often have many different facets. Food security of any kind is impossible without enough food at the right time and in the right place, without adequate transportation networks for delivering food within and among nations, and without enough income to buy imported food when local supplies are inadequate. The reality of global interdependence is vividly illustrated by the way food security in one area can be severely threatened by events halfway around the world.

When local harvests in a developing nation are insufficient to feed the people of that country, families who barely have enough to eat in normal times are forced to eat even less. A dire need also exists for international emergency food reserves to be established that are continually available.

Even though the sheer number of hungry people around the world poses a monumental problem, when consideration is given to the vast land resources that are not being fully utilized for food production, it becomes increasingly evident that the true problem today lies in the inescapably linked factors of overpopulation and underdevelopment. Today, there is more food in the world than ever before; yet most people do not eat better.

Political Instability and Civil and International Conflict

Refugees who have fled their native lands to escape persecution make up a sizeable portion of the world's hungry. These people are usually found in camps in nations bordering their country of origin. Destitute, often ill, and almost always victims of undernutrition and malnutrition, such people must depend on direct food aid from donor organizations. Often the nation in which the refugees have sought temporary asylum finds it difficult to provide the minimum amounts of food and health care needed, much less provide a means for educating the children. Developed nations may send assistance, but when the refugees attempt to immigrate to industrialized countries, controversies arise as to whether they can be admitted as "political refugees" or turned away on the basis of their classification as "economic refugees." Contributing to the devastating chaos so often present in refugee camps is the tactic used by insurgent guerrillas of infiltrating the camps and setting up headquarters within the vicinity. To assist with meeting the nutritional needs of those living in refugee camps, some organizations provide aid through establishment of fields for the growing of food crops. While this does not entirely eliminate the need for direct food aid, it does tend to provide some encouragement to the refugees through their active involvement on their own behalf.

Even those inhabitants who are not physically driven from their country suffer from the diversion to the military sector of resources normally allocated for food production. Expenditures by a number of nations for armed conflict are such that the amount spent in one month would have, if diverted to food production or acquisition, provided adequately for all the inhabitants for over a year.

Persistent High Rates of Population Growth

There is unanimous agreement that continued high rates of population growth gravely threaten any attempt to reach a permanent solution to the world food dilemma.

Highly Disadvantageous Position of Developing Countries in Terms of World Trade and Economic Development

While some disagreement exists as to the nature of the disadvantages suffered by developing nations in the world's trade complex, no one can deny that a high percentage of transactions tend to result in the greater economic benefit accruing to the more developed trader.

Toton (1982) indicated exports are an important source of foreign exchange for the Third World. Unlike rich nations that transact most of their trade among themselves, the bulk of Third World trade (80 percent) flows between themselves and the industrial nations of the world. It is important for the Third World to have access to the markets of the industrialized nations, and to also receive a fair price for its products.

Further complicating the problem is the success of multinational corporations in their growing influence and control over manufacturing and processing of primary products. A major constraint on developing nations' securing a more advantageous position in terms of world trade is the belief that each country must seek to enact tariffs, embargoes, and other legal means to keep a favorable position. Other restrictions that should be loosened are the provisions under which developing nations secure loans from the International Monetary Fund.

High Levels of Illiteracy and Limited Opportunities for Acquisition of Practical Knowledge and Skills

Although some developing nations have made remarkable progress in lowering the proportionate number of illiterate citizens, the low level of educational attainment remains a major constraint on agricultural development and increased production of food crops.

According to a World Bank report (1988), the level of formal schooling of farmers is correlated with their efficiency as farm managers, and exposure of farmers to extension education improves agricultural efficiency. There are also indications that the formal schooling level of rural women is often correlated with the number of children they bear. Changes in agricultural productivity and population growth are two important dimensions of rural development. To the extent that education and adult information services do influence these variables, alternative government education policies may affect the course of rural development.

The contributions needed from higher education are very slow in coming to most Third World countries. The low prestige associated with agricultural education and extension constitutes a strong barrier that tends to prevent needed professionals from becoming effective with farmers.

The International Research Institutes have been hailed as a great boon for achieving substantial increases in food crops and animals produced in the Third World. Some of the Research Institutes, notably the International Rice Research Institute located in the Philippines, also function effectively as training centers for extension workers and other professional agriculturists. Joint cooperative activities should be encouraged between research centers, colleges of agriculture, and ministries of agriculture. The agricultural education and extension programs should receive services from each of these three groups, and incentives should also be provided that will encourage local agricultural professionals to avail themselves of such services as intensive short training courses, field days, and internships at research institutes.

Finally, community- and village-based adult education is beginning slowly but surely to make more food a reality. The PVOs have perhaps been the most successful in developing techniques, organizing programs, and training local leaders.

Reluctance of Third World Governments to Make Policies and Allocate Resources Favorable to Agriculture

Although the importance of enhancing the performance of the agricultural sector is becoming more evident among governments of developing countries, there is still considerable progress to be made. It is becoming realized that an ailing agriculture will hamper industry, as the two are interdependent.

Political stability of a country is important in establishing and achieving national food production goals. Top levels of the government must vigorously support strong financial commitments to adaptive research, training scientists, and a dynamic agricultural education and extension program.

It is plausible for nations to commit the resources needed to implement a workable strategy to produce the food they require for the decades ahead. Some of the elements needed to successfully accelerate national agricultural production include the following:

- production goals must be established by each nation with full support from the government, industry, and the farmers.
- priority should be given to those agricultural commodities that are produced on the largest land area and involve the largest number of farms and farmers. (This is inclusive of the agribusiness infrastructure.)
- technology for each agricultural commodity must be field-tested and proven in the localities where it will be used.
- successful food production campaigns will be successsful through the combined efforts of research, teaching, extension, and government assistance.
- the farmers must be provided with financial incentives to produce more.
- increased agricultural production must be supported with provisions for adequate amounts of high-quality seed and storage. Low-cost credit must be available to the farmers to enable them to use more production inputs so that maximum use is made of all land.

> Tangible evidence that links schooling to economic development has never been found; it is generally assumed, nevertheless, that the problems of poverty and food production can be resolved by appropriate educational programs.
>
> (MacKinnon, 1985)

The food-poverty-illiteracy-population problem is certainly not a myth. It is indeed a reality! Based on projected world population figures for the year 2100, there is great potential that the problem will become even more compounded as a result of increasing numbers of people alone. Although MacKinnon (1985) suggested there is "no tangible evidence" that links schooling to economic development, there is no question that as a person becomes more educated, their standard of living increases, they opt for fewer children, they become consumers of goods and services, and they contribute in a more meaningful way to their society. It becomes very evident that education is one key to a better quality of life. There is not one among us who can honestly refute the need, demand, or value of education. Our citizens want to be able to do things for themselves—our world neighbors are no different.

Helping people help themselves should be the basic underlying philosophical foundation for designing international programs in developing nations (Diamond, 1984). People in developing nations tend to have, in varying degrees, an interwoven ethnic and national pride that is characterized by cultural dignity and integrity. Constantly providing the populace in a developing nation with charitable bounties from more developed nations can ultimately result in destroying these qualities. However, donor nations that design programs aimed at helping people help themselves can reinforce and instill pride, dignity, and integrity.

To achieve this end, three essential functions must be addressed. First is the identification of talents and skills that exist within a society. These talents and skills are already in place, and are being passed on from one generation to another because they have helped society survive for hundreds of years. Second is the identification of natural resources that exist and are readily available, whether they be water, soil, trees, people, animals, grass, rocks, bamboo, or a combination of several. Project goals should be aimed at using these resources instead of relying too much on imported resources. Such natural resources are already an integral part of the culture, are accepted by the populace, and will be available long after the projects funded by donor nations have been phased out. Third is to introduce modern agricultural concepts by using intrinsic talents and skills already in place. Such concepts should address the needs expressed by a nation's populace.

Agricultural technicians, specialists, and educators are capable of transferring modern technology and concepts to people in a way that would reflect relevancy to their expressed needs. The impact should be greater when local skills, talents, and natural resources are manipulated into development project goals of donor nations.

Education cannot be forced upon people; they must have a felt need and look with trust and approval upon those who attempt to give it. Education must address those expressed needs and blend them into cultural mores. It is crucial that modern concepts not compete with or replace traditional practices in a culture, but are added to the culture for consideration, acceptance, and ultimate implementation by those having the expressed need. There are certain essentials that have to occur within any country before developmental assistance can have an impact, and they include the following:

1. there has to be a commitment of the citizens and the government.

2. development must be brought about by knowledgeable agriculturists and practical business people.
3. there needs to be greater incentives that offer potential for jobs and profit.
4. there must exist an opportunity to expand educational programs.
5. there must exist an opportunity to engage in agricultural research.
6. basic roads, transportation, and markets need to be taken into consideration.
7. development of other types (not just education and agriculture) should be encouraged.
8. there should exist a moral and sympathetic leadership.

Obviously, many nations realize the necessity to work cooperatively and have made it a priority to increase agricultural production and farm incomes by implementing agricultural innovations. In an attempt to increase agricultural production and income in Nigeria, the government and institutions of higher learning have been working on the problem of planned agricultural development aimed at revolutionizing agriculture through the introduction of different types of technological innovations (Okuneye, 1985). The main objective of these efforts is to raise the standard of living of subsistence farmers (Atala, 1984). At the same time, Atala (1984) indicated that a variety of factors have inhibited widespread adoption of recommended agricultural innovations, and include educational, economic, political, sociological, and sociopsychological factors. However, the lack of highly trained agricultural scientists and extension specialists seems to be one of the most serious constraints to the development and transfer of improved agricultural technology to the Nigerian farmer (USAID, Africa Bureau, 1984).

China, which certainly has its population problem, also recognizes the need for educational reform and, like Nigeria, is experiencing difficulty in providing a sufficient number of trained agriculturists. As reported by Zhang and Holt (1989), one of the results of educational reform in China since 1978 has been the rapid development of vocational education. By the end of 1986, there were 3,187 vocational and agricultural middle schools enrolling 2.5 million students. There were also 3,782 secondary specialized schools with 1,757 students in the same year. The ambitious goal of the government is to produce 1.1 times more graduates from secondary vocational technical schools during the next five years (1986–1990) than the previous ones.

In rural areas, many secondary schools either have started to offer vocational courses or have been transformed into secondary agricultural-technical schools. These new programs demand a great number of agricultural teachers. At present, there is an acute shortage of teachers for agricultural education. In order to meet this high demand, many agricultural education programs have been established since the early 1980s.

Here in the U.S., the National Research Council in its publication "Understanding Agriculture—New Directions for Education" (1988) asserted the following:

- most Americans know very little about agriculture, its social and economic significance in the United States, and particularly, its links to human health and environmental quality.
- few systematic educational efforts are made to teach or otherwise

develop agricultural literacy in students of any age. Although children are taught something about agriculture, the material seems to be fragmented, frequently outdated, usually farm oriented, and often negative or condescending in tone.

Based on these observations, the National Council recommended, that all students should receive at least some systematic instruction about agriculture beginning in kindergarten or first grade and continuing through twelfth grade. Much of the material could be incorporated into existing courses and would not have to be taught separately.

Currently, the National Council is implementing a project to infuse international agriculture into the agricultural education curriculum at the secondary level. The National Council's rationale for infusing such a program is as follows:

The mission of agricultural education in the United States of America is to foster the development of knowledge and skills to the industry of agriculture. Pursuant to this mission is a growing need for students and educators to develop an understanding of world agriculture and its impact on U.S. agriculture, as well as its effect on local production and marketing of food and fiber. To address this need requires the development of a systematic approach for infusing various aspects of international agricultural systems into the study of agriculture at all levels of the USA agricultural education program.

Specifically, the following major concepts provide the rationale for the development of a systematic approach to the study of international agriculture systems:

1. There is a need for students and educators to develop a global awareness of agriculture.
2. There is a need to understand the importance of agriculture in international trade.
3. There is a need for a better understanding of international agricultural marketing systems, economics, and agricultural production and education systems.
4. There is a need for a better understanding of different cultures and their political systems, and how they relate to the food production and management systems.
5. There is a need for students and educators to be aware of internationally oriented career opportunities.
6. There is a need for students and educators to find ways to enhance problem solving and creative thinking skills.

7. There is a need for students and educators to develop an appreciation for different language skills and develop means to build cultural linkages.
8. There is a need for students and educators to develop an understanding of the interrelationships between education, the scientific support base, and the total food and agricultural system.

Conclusion

Any study of worldwide production, processing, marketing, and distribution of agricultural commodities forces acknowledgement of the overwhelming multiplicity and complexity of the many associated factors. Every nation desires to attain a higher standard of living for its people. Such an accomplishment demands first of all a solidarity of commitment by individual governments. A chief necessity is wise use of both material and human resources. Ample allocation to the agricultural and to the educational sectors is mandatory. Worldwide recognition should be given to the specific needs of developing nations. Much must be done in terms of sharing; this should be done not only in times of emergency, but should be a continual process that contributes to the acquisition of knowledge and skills and that also gives rise to progress in improving the health and well-being of poverty-stricken and disadvantaged people around the globe. The international agricultural commodity complex does assuredly hold great promise for a brighter and better world.

Discussion Questions

1. Describe the world situation relative to food production and consumption.
2. Why are farmers unable to obtain the necessary resources to produce, process, market, and distribute food?
3. Why is the reference to food supplies made such that it leads you to believe worldwide food supplies are insecure? What does the term insecure mean to you?
4. How does civil or international conflict affect food production, processing, marketing, and distribution?
5. Are developing countries at a disadvantage relative to world trade and economic development? If so, how?
6. What contribution does the acquisition of knowledge and skills make toward increased economic development?
7. Why are some governments reluctant to make policies and allocate resources favorable to agriculture?
8. Explain the relationship between food, poverty, population, and illiteracy.
9. What does "helping people help themselves" mean to you?
10. Why should students become more knowledgeable concerning international activities and events?

APPENDIX A
Agricultural Education: Definitions and Implications for International Development

Purpose

The purpose of this manuscript is to discuss the role of agricultural education as "one" component of the agricultural programs of developing nations. Certainly the complexity of agriculture requires an array of inputs (Note: Inputs are of two general types. Institutional support inputs are policies, procedures, and mechanisms which are conducive to agricultural growth. Production-oriented inputs are resources and practices contributing to production enhancement), all of which are important in strengthening the agricultural development process. This manuscript will present rationale for education in agriculture as one of the most crucial of such inputs. To accomplish this objective, the following sections will address (1) the definitions attributed to the term "agricultural education" in the context of international agricultural development, (2) a justification for the inclusion of agricultural education in development programs, and (3) the general role agricultural education could assume in such programs.

Setting the Stage

During the past three decades, international attention has been directed at agricultural production in general, and specifically at the development of the agricultural sectors of the world's lesser developed nations. Agricultural development for these countries is critical. Characteristically, these nations are heavily dependent on agriculture as their primary economic activity (Tinnermeir, 1974; Ryan and Binswanger, 1979). Malassis (1975) indicates that developing nations

---

Reprinted by permission of the NACTA Journal, June, 1984.

typically (1) have extremely high percentages of their populations engaged in agriculture, (2) maintain a high percentage of agricultural exports in relation to total exports, and (3) have agricultural sectors which contribute heavily to the total gross domestic product (GDP). Ironically, however, agricultural GDP per agricultural worker rarely exceeds even half the per capita GDP in these same nations. In short, although developing nations are highly dependent on agriculture, it remains a weak sector of their economies. These conditions point out the urgent need to both stimulate production and enhance the relative stature of agriculture in developing nations.

Definitions of Agricultural Education

A number of individuals have addressed the development process for agriculture, and in so doing have referred to agricultural education. The result is a variety of meanings assigned to the term agricultural education when used in the context of international development.

To those most familiar with the U.S. educational system, agricultural education commonly refers to those activities directed at the preparation of teachers of vocational agriculture. However, this is not the common definition as viewed from the perspective of international development.

For instance, Roberts (1980) infers that international agricultural education consists of programs in higher education such as short-term trainees at U.S. universities, U.S. faculty degree teaching in foreign countries, and short-term training in-country conducted by university faculty members. Broadening this perspective, others classify agricultural education as the general mission of colleges of agriculture in higher education—quality instruction in all agricultural subject areas (Love, 1982).

A popular interpretation of agricultural education is that it is fundamentally synonymous with agricultural extension. This can include either programs for training extension workers or, most commonly, field programs directed at small farmers. Indeed, extension in some form is commonly visualized as the primary mechanism for promoting increased productions via the diffusion of new technologies and their ultimate adoption by farm clientele. Adult education programs for rural populations are a form of extension that are viewed by some as a means for educating adult farmers, both in agriculture and other subject matter areas (Hall and Kidd, 1978). Others advocate the utilization of indigenous knowledge systems and indigenous technology in both adult education and extension. Brokensha, et al. (1980) contend that this methodology greatly enhances the success of extension programs.

Coombs and Ahmed (1974) are also proponents of nonformal extension education as a means for achieving rural development, including agriculture. They define it as any organized, systematic, educational activity carried on outside the framework of the formal system. Thus defined, nonformal education includes, for example, agricultural extension and farmer training programs (p. 8).

The term agricultural education in developing countries has also been used to describe, in aggregate, the various training projects associated with the international research centers located around the world, even though these training efforts are not coordinated among the centers (Rockefeller, 1974, 1976).

Some individuals have suggested definitions for agricultural education that extend beyond those reviewed above. Curle (1970), for example, presents his thesis that education as it relates to agriculture includes the whole range of formal programs that can be initiated by governments. This includes not only extension training, secondary and higher education, but teacher training and vocational education in agriculture as well. Similarly, Kimmel (1982) and Malassis (1975) interpret agricultural education broadly, covering a wide array of possible programs whose purpose is instruction in agriculture.

An all-embracing explanation of agricultural education has also been offered by Habito (1980), who presents an outline for manpower resources in agriculture that encompasses everything from nonformal education to highly technical education at the university level.

It is evident that many definitions have and can be given to agricultural education, especially as it relates to international agricultural development. One must surely conclude that agricultural education consists of any and all organized programs whose purpose is education or training in agricultural subjects.

Justification and Rationale

Previous sections have been directed as setting a proper stage for the topic and reviewing the meaning of international education in agriculture. The next step is to answer the question, "Is there a place for agricultural education in agricultural development?" The response appears obvious. If the development of agriculture is, as has been demonstrated, critical to the economies of developing nations, and if education has a legitimate function to perform in agriculture (a thesis everyone involved in agricultural education in any of its forms could subscribe to), then one can deduce that agricultural education is a legitimate component of agricultural development.

For example, in articulating the need for agricultural education, Kimmel (1982) remarks that according to FAO estimates, in the next 20 years the total number of agricultural extension workers in the world will reach 1.25 million. This is somewhere near four and a half times the current total. The vast majority of these, and other agriculturalists, will have to be trained in their own or neighboring nations. Certainly, a tremendous education effort will be required if these predictions are even to be partially fulfilled.

One must also consider that in most countries in greatest need of agricultural development, almost all arable land is currently under production (Rojko, 1978). The major constraint to increased productivity is the education of producers in the proper use of improved technologies. Indeed, that is the most plausible path to improvement in the agricultural sector. The technologies are currently available. Their innovative adaptation, adoption, and application are the elements to which agricultural education can greatly contribute.

Agricultural education, as defined herein, can, and does, have a positive impact in technology transfer. Empirical studies support this assertion. Shukla's work among small farmers in India led him to conclude that "An effective educational program can do much to shorten the time lag between the discovery of new practice and its adoption by all farmers" (Shukla, 1971, p. 73). Similarly, Moock (1980) utilized an economic production function model to measure the marginal product of education (its effect on the utilization of farm inputs) in a large maize project in Kenya. His research yielded the following conclusions:

> Any form of education that imparts knowledge about the production process directly, or which enhances the capacity to acquire knowledge about the production process from other sources, should raise the individual producer's surface of production possibilities. With any particular combination of inputs, the producer with more production-relevant education can (and will) produce more output (Moock, 1981, p. 738).

These, and other similar observations of the positive effects of education in agricultural projects in developing nations are aptly summarized in the following concept expressed by Ruttan:

Productivity differences in agriculture are increasingly a function of investments in the education of rural people, rather than natural resources endowments. Indeed, the one inescapable implication of the results of our cross country analysis is the importance of literacy and schooling among agricultural producers and of technical and scientific education in the agricultural sciences (Ruttan, 1973, p. 5).

The preceding points clearly both delineate and justify the position that agricultural education has a significant role to play in development. The fundamental rationale is that agricultural sectors in less developed countries can progress only to the extent that the people involved in agriculture progress, and that these people progress significantly only through viable systems of education in agriculture. Granted, other inputs into the sector are also required, but their use and applications are entirely dependent upon the ability of prepared individuals to capitalize upon them. Consequently, the preparation of these people—their education—is crucial since it arms them with the variety of mental and physical skills that efficient modern (as opposed to traditional) agriculture requires.

The Role

While it is clear that there is a role for agricultural education in development, it is difficult to state or to project to what extent that role will be interpreted and manifest by individual nations in agricultural development programs. However, the question can well be asked, "What role should it rightfully have in a balanced development plan?"

Curle (1970, p. 158) advocates that in developing societies, in view of their substantial dependence on agriculture, significant attention should be given to the enhancement of education in agriculture. More specifically, he mentions these actions:

1. To inquire into the incentives needed to induce persons to train as agricultural scientists and other specialists.
2. To study and promote the teaching of science and agriculture at various levels of schooling.
3. To give the cultivators themselves as much education in agriculture as possible, both through extension work and, where feasible, through short courses of instruction.

Although Curle presented these ideas more than a decade ago, they are just as applicable today. More recently, the findings of the Presidential Mission on Agricultural Development in Central America and the Caribbean (Presidential Mission, 1980) suggest a similar course for education in agriculture. To summarize, the Commission's recommendations are to:

1. Expand educational opportunities in agriculture through strong national and regional education and training programs, and through major in-service training for agriculturalists.
2. Strengthen efforts for developing and applying improved technology by initiating major agricultural research programs and improving government extension programs.
3. Elevate the prominence of agriculture as a profession, and to provide opportunities and inducements to young people to study agriculture.
4. Initiate programs to educate adult farmers and farm families both in general literacy skills and the use of modern production techniques.

The preceding comments demonstrate that agricultural education can assume a variety of program roles. They are a clear signal to governments in developing nations of ways in which the spectrum of agricultural education can be significantly enlarged to stimulate and support the agricultural sector. Each nation must determine for itself its role for education in agriculture. But, if education is to be an accelerator of agricultural development, as postulated by Mosher (1966), it must be afforded a sufficient amount of public planning and resources to enable it to truly complement growth (and ignite growth) in agriculture.

Conclusion

Agricultural education, in the context of international development, embraces a wide range of meanings, for it is any organized activity that has as its purpose instruction in agriculture. It is the process by which specialists are produced, agriculturalists are trained, and farmers are assisted. It is, or should be, a partner with other program inputs in the process of development in agriculture. Given the current economic importance of agricultural sectors in developing societies, agricultural education, in its fullest sense, is thus elevated to a role of great importance.

Development in agriculture must begin with people. Their knowledge and skills are the primary input to the human resource. All other inputs are secondary. Progressive agriculture requires capable individuals at all levels, from the policy maker to the farmer and from the researcher to the extension agent, who are skilled in their professions and who understand the intricacies of agriculture. Indeed, the opportunity cost of not sustaining a strong, diversified, and viable system of agricultural education is too high. Of the many constraints inhibiting agricultural development today, perhaps the greatest is the failure to recognize this fact.

Agricultural education can be viewed as a key, one that can unlock long-lasting benefits. Without it, those benefits may be forever "locked-in," never to see the light of day. The hope is that the key will be turned—that it will open a door for developing nations and allow them to improve their contribution to their societies and the world community.

References

Brokensha, D., D. M. Warren, and O. Werner (Editors). INDIGENOUS KNOWLEDGE SYSTEMS AND DEVELOPMENT. Washington, D.C.: University Press of America, Inc., 1980.

Coombs, Philip H. and Manzoor Ahmed. ATTACKING RURAL POVERTY: HOW NONFORMAL EDUCATION CAN HELP. Baltimore: Johns Hopkins University Press, 1974.

Curle, Adam. EDUCATION STRATEGY FOR DEVELOPING SOCIETIES: A STUDY OF EDUCATIONAL AND SOCIAL FACTORS IN RELATION TO ECONOMIC GROWTH. 2nd Edition. London: Travistock Publications, 1970.

Habito, C. P. Suggested Guidelines for Planning and Implementing National Programmes of Agricultural Education and Training. 1979, TRAINING FOR AGRICULTURE AND RURAL DEVELOPMENT (pp. 28–38). Rome: Food and Agriculture Organization of the United Nations, 1980.

Hall, Budd L. and J. Roby Kidd (Editors). ADULT LEARNING: A DESIGN FOR ACTION: A COMPREHENSIVE INTERNATIONAL SURVEY. Oxford: Pergamon Press Ltd., 1978.

Kimmel, D. C. Impact of International Perspectives on American Agriculture in the 80's. NACTA JOURNAL, 26(3), pp. 14–19.

Love, Gene M. Securing America's Food and Agricultural Resource Base. Paper prepared for the Northeast Higher Education Committee, a sub-committee of the Northeast Region Council on Food and Agricultural Science, USDA. 1982.

Malassis, Louis. AGRICULTURE AND THE DEVELOPMENT PROCESS: TENTATIVE GUIDELINES FOR TEACHING. Paris: The UNESCO Press, 1975.

Moock, Peter R. Educational and Technical Efficiency in Small-Farm Production. ECONOMIC DEVELOPMENT AND CULTURAL CHANGE, 29(4); pp. 723–739, 1981.

Mosher, Arthur T. GETTING AGRICULTURE MOVING: ESSENTIALS FOR DEVELOPMENT AND MODERNIZATION. New York: Fredrick A. Prager, Publishers, 1966.

Presidential Mission on Agricultural Development in Central America and the Carribean. AGRICULTURAL DEVELOPMENT AND ECONOMIC PROGRESS IN THE CARRIBEAN BASIN. Tallahassee, Florida: 1980.

Roberts, N. Keith. The Population-Food Squeeze: Education for Survival. Logan, Utah: 62nd Faculty Honor Lecture, Utah State University, 1980.

Rockefeller Foundation. STRATEGIES FOR AGRICULTURAL EDUCATION IN DEVELOPING COUNTRIES. New York: The Rockefeller Foundation, 1974.

Rockefeller Foundation. STRATEGIES FOR AGRICULTURAL EDUCATION IN DEVELOPING COUNTRIES. Second Bellagio Conference, 1975. New York: The Rockefeller Foundation, 1976.

Rojko, Anthony, et al. ALTERNATIVE FUTURES FOR WORLD FOOD IN 1985. U.S. Department of Agriculture, Foreign Agriculture Economic Report No. 146. Washington, D.C.: U.S. Department of Agriculture, 1978.

Ruttan, Vernon W. INDUCED TECHNICAL AND INSTRUCTIONAL CHANGE AND THE FUTURE OF AGRICULTURE. New York: The Agricultural Development Council, Inc., 1973.

Ryan, James G. and Hans P. Binswanger. Socio-economic Constraints to Agricultural Development in the Semi-arid Tropics and ICRISAT's Approach. From Kumble, Vrinda (ed.), PROCEEDINGS OF THE INTERNATIONAL SYMPOSIUM ON DEVELOPMENT AND TRANSFER OF TECHNOLOGY FOR RAINFED AGRICULTURE AND THE SAT FARMER (pp.57–67). Andhra Pradesh, India: ICRISAT, 1979.

Shukla, V. P. Interaction of Technological Change and Irrigation in Determining Farm Resource Use. Jabalpur District, India, 1967–68. CORNELL INTERNATIONAL AGRICULTURAL DEVELOPMENT BULLETIN 20. Ithaca, New York: Cornell University, 1971.

Tinnermeier, Ronald L. Credit for Small Farmers. From H. H. Biggs and Ronald L. Tinnermeier (eds.), SMALL FARM AGRICULTURAL DEVELOPMENT PROBLEMS (pp. 97–116). Fort Collins, Colorado: Colorado State University, 1974.

APPENDIX B
Gaining International Experience

A frequent question asked by many is, "How do I gain international experience?" Fortunately, there are many opportunities for persons to gain international experiences, ranging from short-term exchange programs to volunteer programs to trainee and internship programs to professional assignments. With all of the kinds of opportunities to gain international experience, there should be one that is tailored to each individual's needs.

If gaining experience abroad is educationally meaningful and exciting, gaining experience in the Third World is especially so. The fact that these countries are so different culturally from the U.S. poses a starker contrast with our familiar ways and thus a more striking challenge by which to understand our own society, values, assumptions, and larger world role. While living in a traditional European setting exposes one to new sights and smells, new customs and unknowns, living in Third World countries provides an even greater range of experience. Only by understanding the variations among the world's peoples—their traditions, values, and aspirations—can we perceive the common humanity that unites all societies.

An understanding of Third World settings is valuable preparation for living and working in an increasingly interdependent world. The fact is that over three-quarters of the world's population lives in Third World countries, yet barely ten percent of Americans who study abroad go there. U.S. economic dependence on Third World countries (as well as theirs on us) is extensive and growing—a fact that affects U.S. trade, jobs, and living standards. The Third World also vitally affects U.S. national security; in fact, all of our country's post-World War II military involvements have been in this region. Beyond business and government, our contacts with the Third World through international organizations, tourism, and humanitarian and cultural exchanges are also increasing, with resultant implications for career opportunities. Indeed, the U.S. is itself increasingly populated with Third World people who are enriching our culture and changing the very fabric of our society in ways we need to better understand.

In important cultural and humanitarian terms, our society has much to learn from and contribute to the societies of Asia, Africa, and Latin America. The educated person of the 21st century must be academically and experientially aware of our planet's population in the Third World. Nonetheless, it could be that the likes of gaining experience in the First World (or an industrialized nation) is much more appealing. Rest assured that there are ample opportunities for that as well.

James E. Diamond, in an article entitled, "An Overview: Educational Consultants in International Agriculture," which was published in the NACTA Journal (1986), stated the following:

> Consultant assignments can be operative, challenging, educational, and can have lasting impact when contracted agricultural educators who participate in international assignments generally contribute to the overall educational programs at their respective institutions by being open-minded, tolerant, worldly, and humanitarian. Their reasoning, thinking, and dialectical views tend to be broader in scope as they advise students, teach classes, plan programs, serve research, and perform other important academic responsibilities. Furthermore, they can learn to appreciate and understand the characteristics of other cultures; and, learn to speak a foreign language. Hence, international assignments can have a two-way impact on the understanding of people and societies, both domestically and internationally. The words of the late Eleanor Roosevelt best summarize this concept: "Understanding is a two-way street."

Some Considerations for International Assignments

Persons contemplating entry employment into an international agricultural program should first make a self-appraisal of their attitudes toward international work. An interest in traveling and living in foreign countries and working with rural poor is essential to the development of a positive work attitude toward international projects.

Although many counterparts in international work may speak English, it is essential that an attempt to speak, read, and write the native language of the developing country be made. In developing countries, the simple transfer of existing U.S. programs cannot occur with the expectations of the same or equal results. Modifications must be made to account for sociocultural aspects and the current status of development within the country.

The bureaucratic process involved to acquire approval or actions is a lengthy one in most assistance agencies. Possession of a high degree of tolerance for time-consuming processes that usually involve a number of foreign and domestic governmental agencies is necessary.

Acquisition of supplies and equipment is very slow in developing countries and transportation difficulties can be expected. Flexibility, adaptability, and innovativeness are essential traits for successful international employment. It is very important for consultants and administrators to make the necessary changes to achieve project objectives. (Please refer to Appendix B and review "How to be an Effective Consultant on International Development Projects".)

Gaining Access to Opportunities in International Agricultural Education, Extension, and Research

Preparation for and success in international employment are complex multi-dimensional tasks. The following are suggestions for anyone beginning the process:

- develop some expertise in a technological field that is internationally marketable (agriculture is an excellent choice).
- acquire proficiency in those behavioral sciences that are related to the learning, communicative, and human development processes.
- have some understanding of and appreciation for administrative (bureaucratic) procedures.
- be able to apply all the previous factors to a complex sociocultural setting.

Fortunately, there are many avenues of approach to acquiring these qualifications. How one gains access to international opportunities will be influenced by their current level of professional development.

Suggestions and Advice

1. Learn to speak, read, and write at least one useful language. Currently, French and Spanish are most in demand for qualifying to work in the developing countries.
2. Travel to developing countries to observe ongoing programs and, if possible, to serve a brief internship with an international project. Also, take extended vacations outside the U.S., preferably in developing countries. Mexico is just across the border from the U.S.
3. Attend international meetings. Involvement at international seminars with individuals from developing countries can further one's understanding of the international setting and its problems, as well as enhance one's professional development.
4. Take part in local international activities and programs. This can heighten one's understanding of cross-cultural similarities and differences and, at the same time, lead to information about overseas employment opportunities.

Suggestions For Gaining International Experience

The selected suggestions listed are merely some examples of how a person can become involved and gain valuable international experience. There are many other programs and examples that could be shared.

Teacher Education and the Peace Corps

Over one-third of Peace Corps volunteers are teachers. Since the Peace Corps first sent volunteers overseas in 1961, education has been its largest program. Today, volunteers teach in primary, secondary, and university classrooms in 56 countries ranging from Antigua to Zaire. Subjects include math, science, English, and vocational education. Volunteers also work as instructors in teacher training institutions and serve as resource specialists organizing workshops and conducting on-site observations.

With the countries of the developing world experiencing unparalleled population growth, preparation of teachers is one of the most important tasks facing national governments today. All face severe shortages of trained teachers and educational materials.

To meet this burgeoning need, the Peace Corps is shifting its program focus from individual classroom instruction to teacher education. More emphasis is being placed on the formal transfer of skills to host country teachers, although many volunteers will continue to teach in classrooms. In collaboration with Ministries of Education, volunteers are needed to help untrained teachers improve their skills, develop instructional materials and methodologies, and identify local resources.

Agriculture and the Peace Corps

Being a Peace Corps Volunteer in agriculture will provide the following:

1. The opportunity to help improve the critical agricultural and economic needs of the people in the developing world. Increased food production is the first priority for the poorest nations. More than one fifth of the developing world—an estimated 600 million people—are hungry or starving. Moreover, it is projected that the developing countries of the world will be forced to import up to 50 million tons of food, dragging them further down into debt they cannot repay—unless they receive assistance in modern agro-technologies from Peace Corps volunteers.

2. The opportunity to gain skills in a broad range of agricultural techniques. For example, Peace Corps volunteers may be asked to do the following:

   - introduce plowing by oxen in Togo to replace the old short-handled hoe.
   - help Indians raise strawberries in Guatemala and market them in the U.S.
   - provide technical assistance to beekeepers in Micronesia to improve the bee colonies for profitable honey harvests.
   - set up a demonstration farm at a rural Liberian elementary or junior high school.
   - conduct research in Tonga on the diseases of banana, vanilla, kava, citrus, and root crops.
   - act as an extension agent for beef cattle cooperatives in Kenya.
   - integrate rural Costa Rican women into 4-H club projects to increase their earning potentials.
   - assist in insect-control on a Caribbean island.
   - improve agricultural practices as a crop extensionist in Paraguay.

3. The opportunity to develop communications and leadership skills.

4. The opportunity to use your experience in agriculture where it is needed the most. Crop and dairy farmers, cattle ranches, agronomists, beekeepers, and men and women with mechanical and marketing skills are desperately needed to modernize and improve age-old agricultural systems that are preventing many developing nations from being able to feed themselves. Help a few farmers reap more bountiful harvests and you will help their country grow.

5. The opportunity to broaden horizons and background for a successful career in agriculture.

Employers recognize and highly regard the work experience of Peace Corps volunteers. Their credentials are right on target to qualify for many positions in the international area, where overseas experience and language ability are enormous assets. Or those experiences can translate into a more interesting, challenging position within the U.S., because employers in the U.S. also value and seek out skills gained through Peace Corps service.

The Peace Corps is looking for people with backgrounds in the following:

- plant protection
- crop development
- soil science
- agriculture education
- animal husbandry
- community agriculture
- farm mechanics
- beekeeping
- rural youth projects

Work Experience Abroad (Work Exchange Program of the National FFA Organization)

The Work Experience Abroad (WEA) program is designed for active and alumni FFA members ages 18 to 24 (16 to 19 for the Japan Short Program). The program's goal is to offer practical work experience in another country and provide for the observation and study of agricultural methods. Students also become familiar with the history and culture of the people of another country. Participants become representative of the U.S. as they participate in activities to exchange ideas and improve agriculture and rural youth organizations in another country.

In order to participate in the WEA program, an applicant must meet the age requirements, have good practical experience in farming, ranching, horticulture, or other specialized fields; and, be recommended by the high school agricultural education teacher and State FFA Advisor. Applicants going to countries with a language other than English should have a sincere desire to learn a new language.

Participants should be willing to research agriculture in their home state, plus have a basic knowledge of American agriculture on a broad basis in order to receive the most benefit from the program by sharing their knowledge with their host family.

Programs vary in length from two to six weeks on special short exchanges up to 12 months. Twelve-month Around The World (ATW) participants have the wonderful opportunity to see agriculture on three continents.

4-H International (United States Department of Agriculture)

4-H, the youth organization of the Cooperative Extension Service, has been active in exchange programs for decades. For example, since the Peace Corps began in 1961, 4-H has been a viable partner in 60 countries currently served by the 4-H/Peace Corps programs. For more information, contact your local County Cooperative Extension Office.

Students Work Abroad (Council on International Educational Exchange)

The Council on International Educational Exchange is a private, not-for-profit membership organization incorporated in the U.S., with international offices, affiliations, and representation. Major responsibilities and contributions to the field of educational exchange, throughout its four decades of service to the academic community, have established the Council as one of the foremost organizations concerned with international education and student travel.

The Council was founded in 1947 by a small group of organizations active in the field of international education and student travel in order to help reestablish student exchange after World War II. In its early years, the Council chartered ocean liners for transatlantic student sailings, arranged group air travel, and organized orientation programs to prepare students and teachers for their educational experiences abroad. Over the years, the Council's mandate broadened dramatically with the ever-increasing number of its academic members whose interests spread beyond Europe to Africa, Asia, Latin America, and Oceania. Today, the Council assumes important educational responsibilities and develops programs of international educational exchange throughout the world on behalf of its North American and international constituencies.

Congress-Bundestag Youth Exchange Program (a Full Scholarship Exchange Program for American and German High School Students)

Conceived and funded by members of the U.S. Congress and German Bundestag, the program is designed to strengthen ties between the new generations in each country. The program enables the exchange of students, who will go on to shape our society in the future, to expand their perspectives and awareness of German and American social, economic, and political institutions, while extending friendships across international boundaries.

Agency for International Development (AID)—International Development Intern Program

AID administers foreign economic assistance programs of the U.S. Government. It operates from headquarters in Washington, D.C. through field missions and representatives in developing countries in Africa, Asia, Latin America, the Caribbean, and the Near East. AID's purpose is to help people in the developing world acquire knowledge and resources to build the economic, political, and social institutions needed to maintain national development. Such assistance covers many diverse areas including agriculture, food and nutrition, family planning and health, education and human resources, energy, environment, natural resources, and private enterprise.

Educational requirements include a graduate degree in agriculture, agricultural economics, economics, or international relations. Current salaries for interns, depending on education and work experience, are in the $19,000 to $25,000 range, and are supplemented by standard Foreign Service allowances when stationed overseas.

World Teach

World Teach, Harvard University, invites college graduates to spend a year as a volunteer teacher in Costa Rica. No prior teaching experience or knowledge of Spanish is required. The school provides housing and a modest salary. It is an opportunity to experience Latin American culture.

Practically every major university in the U.S. has an office of International Programs, and they provide assistance in arranging for tours, study abroad programs, exchange programs, internships, and employment opportunities.

Sunday Newspaper

Listed in the advertisement section of Sunday and daily newspapers are announcements pertaining to tours, study abroad, and employment.

Other Publications That May be of Assistance

1. Basic Facts on Study Abroad, 1987. Council of International Educational Exchange, Institute of International Education, and the National Association of Foreign Student Affairs (Free).
2. Work, Study, Travel Abroad: The Whole World Handbook. CIEE. New York: St. Martin's Press, 1988–89. A complete guide to work, study, travel abroad; includes 800+ study abroad options, advice on independent study, overseas employment, voluntary service, and budget travel ($8.95).
3. Volunteer! The Comprehensive Guide to Voluntary Services in the U.S. and Abroad. CIEE and the Commission on Voluntary Service and Action. Describes over 170 voluntary service opportunities worldwide ($4.95).
4. Academic Year Abroad, 1988–89. New York: Institute of International Education. Annual. Describes over 1,200 programs worldwide offered by accredited U.S. colleges and universities. Available from IIE, 809 UN Plaza, New York, NY 10017 ($19.95).
5. Vocational Study Abroad, 1988. New York: Institute of International Education. Annual. Hundreds of study abroad opportunities around the world (same address as Academic Year Abroad) ($19.95).
6. Directory of International Internships, 1987. Lists 500 international internships offered by educational institutions, government agencies, and private organizations. Available from the Office of Overseas Study, 108 International Center, Michigan State University, East Lansing, MI 48824-1035 ($12.50).

Some of the largest international corporations are American, and many offer overseas assignments. Fortune Magazine's listing of the World's Biggest Industrial Corporations rates General Motors, Ford Motor Company, and Exxon as the three largest world firms, with IBM as No. 5, General Electric as No. 6, and Mobil as No. 8.

Most international firms want their executives to have some overseas experience. Merck has internationalized its training program and Dow Chemical wants future CEOs with experience running foreign operations. At least 80 percent of the top three hundred executives at Xerox Corporation have had international experience. Key to applying with these firms is to apply as a skilled expert with international expertise.

There also are international opportunities in various professions. The same rule of thumb applies—specialize in a professional or business area along with an international area.

U.S. Government Organizations

Agency for International Development
Office of Personnel and Management
320 21st Street, NW
Washington, DC 20523
202/632-9608

Arms Control and Disarmament Agency
Department of State Building
Washington, DC 20451
202/647-4000

Central Intelligence Agency
Recruitment Office
Ames Center Building
1820 North Fort Myer Drive
Arlington, VA 22209
703/351-2028

Central Intelligence Agency
Director of Personnel
Washington, DC 20505
703/351-2028

Department of Defense Dependents School
Teacher Recruitment Section
Hoffman Building I
2461 Eisenhower Avenue
Alexandria, VA 22331
(enrolls about 140,000 students abroad for military dependents; 10,000 positions, most school positions available.)

Defense Intelligence Agency
The Pentagon
Washington, DC 20301
202/694-4780

Defense Security Assistance Agency
The Pentagon
Washington, DC 20301
202/695-5931

Foreign Broadcast Information Service
Personnel Office
P.O. Box 2604
Washington, DC 20013

International Development Cooperation Agency (IDCA)
320 21st Street, NW
Washington, DC 20523
202/632-3348
(supervisory agency to AID, OPIC, and Trade and Development Program.)

National Security Agency
Attention: Office of Employment
Fort Meade, MD 20755
301/688-6311
(principally concerned with signal intelligence, codes, computer security, and foreign intelligence. Hires a large number of linguists, technicians, and computer experts. Most applicants must take professional qualification test, usually in late November. Write for bulletin.)

National Security Council
Executive Secretary
Executive Office Building
Washington, DC 20506
202/395-3044
(advises the president on security issues; extremely small expert staff, almost no hiring.)

Office of International Health
Public Health Service
5600 Fishers Lane
Rockville, MD 20857
301/443-5460

Overseas Private Investment Corporation (OPIC)
1129 20th Street, NW
Washington, DC 20527
202/653-2807
(international financing and insurance, personnel office has career brochure; over 130 employees.)

Peace Corps Volunteer Recruitment Program
Peace Corps

Office of Recruitment, Room P-301
806 Connecticut Avenue, NW
Washington, DC 20526
202/254-9814

Peace Corps Staff Recruitment Information
Peace Corps
Office of Personnel, Room P-304
806 Connecticut Avenue, NW
Washington, DC 20526
202/254-3400 or 800/424-8580

Personnel Division
Foreign Agricultural Service
(U.S. Department of Agriculture)
South Building, Room 5627
Washington, DC 20250
202/382-1587

Personnel Office
Office of International Cooperation and
    Development
Auditors Building, Room 3118
U.S. Department of Agriculture
Washington, DC 20250
202/475-4071

U.S. Department of Commerce
14th Street and Constitution Avenue, NW
Washington, DC 20230
202/377-4807 or 202/337-4717
    (FCS personnel)

U.S. Department of Energy
Office of the Assistant
Secretary for International Affairs and Energy
    Emergencies
Forrestal Building
1000 Independence Avenue, SW
Washington, DC 20585
202/252-8731

U.S. Department of Labor Bureau of
    International Labor Affairs
OASAM
Office of Operating Personnel Services, Room
    C-5512
Washington, DC 20210
202/523-6717

U.S. Department of State
Foreign Service Officer
Recruitment Branch
P.O. Box 9317
Rosslyn Station
Arlington, VA 22209
703/235-9392

Employment Information Office
U.S. Department of State, Room 2815
Washington, DC 20520
202/632-0580

Language Services Division
U.S. Department of State
Washington, DC 20502 (for translators)

U.S. Department of Transportation
Office of International Policy and Programs
Room 10300
400 Seventh St., SW
Washington, DC 20590
202/426-9630

U.S. Department of the Treasury
Office of the Assistant Secretary for
    International Affairs
Main Treasury, Room 3430
Washington, DC 20220
202/566-5363

U.S. Environmental Protection Agency
Office of International Activities, A-106
401 M Street, SW
Washington, DC 20460
202/382-2973

U.S. Information Agency
1750 Pennsylvania Avenue, NW
Washington, DC 20547
202/485-2618

U.S. International Trade Commission
701 E Street, NW
Washington, DC 20436
202/523-0182

Voice of America
330 Independence Avenue, SW
Washington, DC 20547
202/472-6909

Foundations/Private Organizations

ACCCION International
1385 Cambridge Street
Cambridge, MA 02139
617/492-4930
(sponsors limited capital small business projects in South America; workshops.)

African-American Institute
833 United Nations Plaza
New York, NY 10017
212/949-5666

American Committee on Africa
198 Broadway
New York, NY 10038
212/962-1210

American Council on Germany
14 East 60th Street
Suite 606
New York, NY 10022
212/826-3636

American Friends Service Committee
Recruitment
1501 Cherry Street
Philadelphia, PA 19102
215/241-7000
(specializes in grassroots community development, food aid, legal aid, and disarmament; regional U.S. centers and overseas operations.)

American Fund for Czechoslovak Refugees, Inc.
1776 Broadway
New York, NY 10019
212/265-1919

American Jewish Committee
165 East 56th Street
New York, NY 10022
212/751-4000

American Medical and Research Foundation
420 Lexington Avenue
New York, NY 10170
212/986-1835
(health care to Africa; health professionals required.)

American Near East Refugee Aid
President
1522 K Street, NW; No. 202
Washington, DC 20005
202/347-2558

American Refugee Committee
Director, International Programs
2344 Nicollet Avenue, Suite 350
Minneapolis, MN 55404
612/872-7060
(medical personnel; primarily volunteers.)

AMIDEAST (American Friends of the Middle East)
American-Mideast Educational and Training Services
1717 Massachusetts Avenue, NW
Washington, DC 20036
202/785-0022

Brookings Institution
1775 Massachusetts Avenue, NW
Washington, DC 20036
202/797-6000
(premier research organization; one program is Foreign Policy Studies; over two hundred staff.)

CARE, Inc.
660 First Avenue
New York, NY 10016
212/686-3110
(one of the largest international/technical assistance organizations, hundreds of employees; experience in developing countries, degree in appropriate field preferred.)

Carnegie Endowment for International Peace
11 Dupont Circle, NW
Washington, DC 20036
202/797-6400
(published Foreign Policy magazine, about one hundred employees.)

Catholic Relief Services
Manager of Staffing/Recruitment
1011 First Avenue
New York, NY 10022
(Catholic development, relief agency in over sixty countries.)

Center for Inter-American Relations
680 Park Avenue
New York, NY 10021
212/249-8950

Center for Strategic and International Studies
Personnel Director
1800 K Street, NW; Suite 400
Washington, DC 20006
202/887-0200
(public policy research institute affiliated with Georgetown University; internships available.)

Chicago Council on Foreign Relations
Vice-President and Program Director
116 South Michigan Avenue
Chicago, IL 60603
312/726-3860

China Institute in America
125 East 65th Street
New York, NY 10021
212/744-8181

Christian Children's Fund
203 East Cary Street
Richmond, VA 23251
804/644-4654

Church World Service
475 Riverside Drive
New York, NY 10115-0050
212/870-2257

Council on Economic Priorities
Administrative Director
30 Irving Place
New York, NY 10003
212/420-1133

The Council on Foreign Relations
Personnel Manager
58 East 68th Street
New York, NY 10021
212/734-0400
(published Foreign Affairs; about one hundred employees, mostly Ph.D.'s; regional committees.)

Ford Foundation
320 East 43rd Street
New York, NY 10017
212/573-5000

Foreign Policy Association
205 Lexington Avenue
New York, NY 10016
212/481-8450
(sponsors educational programs on television.)

Hudson Institute
Vice-President, Finance and Administration
620 Union Drive
P.O. Box 648
Indianapolis, IN 46206
317/632-1787
(famous policy research organization.)

Institute of International Education
Personnel Office
809 United Nations Plaza
New York, NY 10017
212/883-8200
(largest exchange education agency in U.S.; regional U.S. and international offices, about 150 employees.)

Inter-American Foundation
1515 Wilson Boulevard
Rosslyn, VA 22209
703/841-3868
(promotes development in Latin America.)

International Human Assistance Program
Recruitment, Program Development
360 Park Avenue South
New York, NY 10010
212/684-6804
(community development, three years experience preferred; French, Spanish, Arabic.)

International Institute for Environment and Development
Deputy Director
1717 Massachusetts Avenue, NW; Suite 302
Washington, DC 20036
202/462-0917

International Research and Exchanges Board
655 Third Avenue
New York, NY 10017
212/490-2002
(sponsors U.S.-Soviet, Eastern Europe academic exchanges.)

Meals for Millions/Freedom from Hunger Foundation
1664 DaVinci Court
P.O. Box 2000
Davis, CA 95616
916/758-6200

Meridian House International
1630 Crescent Place, NW
Washington, DC 20009
202/667-6800

Middle East Institute
1761 N Street, NW
Washington, DC 20036
202/785-1141

Nature Conservancy International Program
Director of Administration
1785 Massachusetts Avenue, NW
Washington, DC 20036
202/483-0231

Near East Foundation
29 Broadway, Suite 1125
New York, NY 10006
212/269-0600
(technical assistance, principally agricultural in Africa, Mideast.)

Overseas Development Council
1717 Massachusetts Avenue, NW
Washington, DC 20036
202/234-8701

Overseas Education Fund
2101 L Street, NW; Suite 916
Washington, DC 20037
202/466-3430

International League for Human Rights
432 Park Avenue South
New York, NY 10017
212/684-1221

International Rescue Committee
386 Park Avenue South
New York, NY 10016
212/679-0010
(aids in counseling, resettling refugees.)

International Voluntary Services
1424 16th Street, NW
Washington, DC 20036
202/387-5533
(small scale development projects; two-year volunteers recruited.)

Oxfam-America
115 Broadway
Boston, MA 02116
617/246-3304
(sponsors integrated rural development, food aid in poorest parts of the world; about fifty U.S. employees, hundreds abroad.)

Partnership for Productivity International
2001 S Street, NW
Washington, DC 20009
202/234-0340
(promotes entrepreneurship abroad.)

Pearl S. Buck Foundation
P.O. Box 181
Perkasie, PA 18944
215/249-0100
(aids Amerasian children; about two hundred employees.)

Project HOPE
People to People Health Foundation
Millwood, VA 22646
(about 150 program staff, many volunteers; primarily health care; graduate degree required.)

RAND Corp.
1700 Main Street
Santa Monica, CA 90406
213/393-0411
(premier research institute; much national security, public welfare research, over five hundred professionals employed, many Ph.D.s.)

Rockefeller Foundation
1133 Avenue of the Americas
New York, NY 10036
212/869-8500
(awards grants, administers field programs abroad; over one hundred employees.)

Save the Children Federation, Inc.
54 Wilton Road
Westport, CT 06880
203/226-7271

SRI International
333 Ravenswood Avenue
Menlo Park, CA 94025
415/326-6200
(consulting and technical research on contract for U.S. government and international agencies; one of the premier think tanks in the world, over three thousand employees, many Ph.D.s.)

Tinker Foundation
645 Madison Avenue
New York, NY 10022
212/421-6858
(promotes better relations between U.S. and Spanish, Portuguese-speaking countries, Ph.D.s preferred.)

Trilateral Commission
345 East 46th Street
New York, NY 10017
212/661-1180
(promotes U.S.-Euro-Japanese cooperation; Ph.D.s preferred.)

Tolstoy Foundation
Executive Director
200 Park Avenue South
New York, NY 10003
212/677-7770
(assists refugees, offices in U.S. and abroad.)

Winrock International
Route 3 Petit Jean Mountain
Morriltown, AR 72110-9537
501/727-5435
(agricultural development, technical assistance; about one hundred employees in U.S. and abroad.)

World Policy Institute
777 U.N. Plaza
New York, NY 10017
212/490-0010
(small policy organization)

YMCA of the USA
Personnel Records Manager
101 North Wacker Drive
Chicago, IL 60606
312/269-0505
(YMCA International offers technical assistance abroad; about fifty employees.)

International Organizations

▶ **NOTE** When possible, U.S. addresses have been listed.

Asian Development Bank
2330 Roxas Boulevard
Pasay City, Phillipines

Food and Agriculture Organization, United Nations
Liaison Office for North America
10011 22nd Street, NW
Washington, DC 20437
202/653-2402

General Agreement on Tariffs and Trade
Center William Rappard
154 Rue de Lausanne
CH-1211 Geneva 21
Switzerland

Inter-American Development Bank
Recruitment Office
808 17th Street, NW
Washington, DC 20577
202/623-1000

International Bank for Reconstruction and Development
World Bank
1818 H Street, NW
Washington, DC 20433
202/477-1234

International Labor Organization
Washington Branch
1750 New York Avenue, NW
Washington, DC 20006
202/376-2315

International Finance Corporation
1818 H Street, NW
Washington, DC 20433
202/477-1234
(an affiliate of the World Bank, promotes private enterprise in developing countries.)

International Civil Aviation Organization, United Nations
P.O. Box 400
1000 Sherbrooke Street West
Montreal, H3A 2R2 Canada

International Fund for Agricultural Development
United Nations
New York, NY 10017

International Monetary Fund
700 19th Street, NW
Washington, DC 20431
202/623-7000

International Telecommunication Union
Place des Nations
Geneva, Switzerland
(4122)99-51-11

Bureau for International Communications and Information Policy:
Call: 202/647-2592

Organization of American States
17th Street and Constitution Avenue, NW
Washington, DC 20006
202/458-3000

Organization for Economic Cooperation and Development
200 L Street, NW; Suite 700
Washington, DC 20036-4095
202/785-6323
(headquarters is in France)

Pan American Health Organization
Pan American Sanitary Bureau
525 23rd Street, NW
Washington, DC 20037
202/861-3200
(coordinates health activities in the Americas.)

South Pacific Commission
U.S. Liaison: Department of State
Office for Pacific Island Affairs
Washington, DC 20520
202/647-3546
(advises on Pacific development.)

United Nations
New York, NY 10017
212/754-1234
for career information:
U.N. Recruitment Staff
Bureau of International Organization Affairs
Department of State, Washington, DC 20520
or:
Recruitment Programs Section
Office of Personnel Services
United Nations
New York, NY 10017

United Nations Children's Fund (UNICEF)
866 U.N. Plaza
New York, NY 10017
212/326-7000

U.N. Conference on Trade and Development
Palais des Nations
CH-1121 Geneva 10
Switzerland

United Nations Development Program
Division of Personnel
Recruitment Section
One U.N. Plaza
New York, NY 10017
212/906-5000

United Nations Educational, Scientific, and Cultural Organization
UNESCO Liaison Office
United Nations
New York, NY 10017
212/963-5995

United Nations High Commissioner for Refugees
UNHCR Liaison Office
1718 Connecticut Avenue, NW
Washington, DC 20036
202/387-8546

U.N. Relief and Works Agency for Palestine Refugees in the Near East
United Nations Secretariat
New York, NY 10017
212/963-1234

United Nations University
Toho Sheimei Building
15-1 Shibuya 2-Chome,
Shibuya-ku
Tokyo, 150 Japan

United Nations Volunteers
United Nations Development Program
1889 F Street, NW; Ground Floor
Washington, DC 20006
202/289-8674
(U.N. equivalent of the Peace Corps. U.S. citizens may also apply through the Peace Corps.)

International Associations

Academy of International Business
Cleveland State University
World Trade Education Center
Cleveland, OH 44115
216/687-3733

American Society of International Executives
International Group Limited
Dublin Hall, Suite 419
Blue Bell, PA 19422
215/643-3040

International Economic Policy Association
1400 I Street, NW
Washington, DC 20005
202/898-2020

International Executive Service Corps
P.O. Box 10005
Stamford, CT 06904
203/967-6000

National Foreign Trade Council
100 East 42nd Street
New York, NY 10017
212/867-4185

International Directories

▶ **NOTE** International chamber of commerce directories list companies engaged in bilateral trade and investment.

American Chamber of Commerce in Ireland
20 College Green
Dublin 2, Ireland

American Chamber of Commerce in Argentina
Avenida R. Saenz Pena, 567
AR-1352
Buenos Aires, Argentina

American Chamber of Commerce in Austria
Tuerkenstrasse 9
A-1090
Vienna, Austria

American Chamber of Commerce in Chile
669 Huerfanos, Office 614
Santiago, Chile

American Chamber of Commerce in France
21 Avenue George V
F-75008
Paris, France

American Chamber of Commerce in Germany
666 5th Avenue
New York, NY 10103
212/974-8830

American Chamber of Commerce in Hong Kong
Swire House, Room 1030
Hong Kong, Hong Kong

American Export Register
Thomas International Publishing Co.
One Penn Plaza
250 West 34th Street
New York, NY 10019

National Council for U.S.-China Trade
1818 N Street, NW
Washington, DC 20036

American-Hellenic Chamber of Commerce
960 Avenue of the Americas
New York, NY 10001
212/629-6380

American Chamber of Commerce (UK)
75 Brook Street
London W1Y 2EB
United Kingdom

Association of Asian-American Chambers of Commerce
1625 K Street, NW
Washington, DC 20006
202/638-1764

Brazilian-American Chamber of Commerce
22 West 48th Street, Suite 404
New York, NY 10036
212/575-9030

British-American Chamber of Commerce
275 Madison Avenue
New York, NY 10016
212/889-0680

Chamber of Commerce Uruguay-U.S.A.
Bartolome Mitre 1337, Esc. 108
Montevideo, Uruguay

Education Information Service
P.O. Box 662D
Newton, MA 02162
617/237-0887
(current openings in education abroad or foreign faculty and administrative openings.)

U.S.-Austrian Chamber of Commerce
165 West 46th Street
New York, NY 10036
212/819-0117

Directory of American Firms Operating in Foreign Countries
World Trade Academy Press
50 East 42nd Street
New York, NY 10017
212/697-4999

Directory of European Industrial and Trade Associations
Gale Research Co.
Book Tower
Detroit, MI 48226
800/233-GALE or 313/961-2242

Directory of Foreign Firms Operating in the United States
World Trade Academy Press
50 East 42nd Street
New York, NY 10017
212/697-4999

Directory of International Trade
Produce Marketing Association
1500 Casho Mill Road
Newark, DE 19714
302/738-7100
(lists nearly one thousand produce exporters and importers in over fifty countries.)

Directory of the U.S.-Arab Chamber of Commerce
One World Trade Center
New York, NY 10048
212/432-0655
(firms involved in U.S.-Arab trade.)

Dun's Latin America's Top 25,000
Dun's Marketing Services
Dun & Bradstreet, Inc.
Three Century Drive
Parsippany, NJ 07054
201/455-0900

Encyclopedia of Geographic Information Sources: International Volume
Gale Research Co.
Book Tower
Detroit, MI 48226
800/233-GALE or 313/961-2242
(lists business information sources worldwide.)

Indo-American Chamber of Commerce
19 South LaSalle Street
Chicago, IL 60603
312/621-1200

Institute of International Education
809 United Nations Plaza
New York, NY 10017
212/883-8200
(publishes sources for jobs overseas.)

International Directory of Business Information
Agencies and Services
Gale Research Co.
Book Tower
Detroit, MI 48226
800/233-GALE or 313/961-2242

International Trade Names Dictionary:
  Company Index
Gale Research Co.
Book Tower
Detroit, MI 48226
800/223-4253
(lists ten thousand firms involved with overseas product marketing.)

Italy-America Chamber of Commerce
350 5th Avenue
New York, NY 10118
212/279-5520

Japanese Chamber of Commerce of New York
115 East 57th Street
New York, NY 10022
212/935-0303

Japanese Chamber of Commerce of Southern California
244 San Pedro St.
Los Angeles, CA 90071
213/626-3067

Korean-American Chamber of Commerce
25 West 32nd Street
New York, NY 10012
212/659-4045

Latin America Market Guide
Dun & Bradstreet International
One Exchange Place, Suite 715
Jersey City, NJ 07302

Latin Chamber of Commerce
1417 West Flagler St.
Miami, FL 33135
305/642-3870

Mexico-U.S. Chamber of Commerce
1900 L Street, NW
Washington, DC 20036
202/296-5198

French-American Chamber of Commerce in the United States
509 Madison Avenue, Suite 1900
New York, NY 10022
212/371-4466

Moody's International Manual
Moody's Investors Service
Dun & Bradstreet Co.
99 Church Street
New York, NY 10007
212/553-0300

Norwegian-American Chamber of Commerce
800 Third Avenue
New York, NY 10022
212/421-9210

Teacher Exchange Branch
United States Information Agency
301 Fourth Street, SW
Washington, DC 20547
202/485-2556
(opportunities abroad for educators)

U.S. Department of Defense—Dependents Schools
Attn.: Recruitment and Assignment Section
2461 Eisenhower Avenue
Alexandria, VA 22331
(overseas employment opportunities for educators)

Philippine-American Chamber of Commerce
711 3rd Avenue
New York, NY 10017
212/972-9326

The World Marketing Directory
Dun & Bradstreet, Inc.
Three Century Drive
Parsippany, NJ 07054
201/455-0900
(50,000 major firms around the world.)

Spain-United States Chamber of Commerce
350 Fifth Avenue
New York, NY 10018
212/967-2170

Swedish-American Chamber of Commerce
825 Third Avenue
New York, NY 10022
212/838-5530

Institute of International Education
809 United Nations Plaza
New York, NY 10017
(covers teaching/educational administrative positions overseas; included information resources.)

Venezuelan-American Association of the U.S.
150 Nassau Street
New York, NY 10038
212/233-7776

Ward's Business Directory of 15,000 Major International Corporations
Information Access Co.
11 Davis Drive
Belmont, CA 94002
415/591-2333

Worldwide Business Directory of the 10,000 Top International Corporations
New Front/Business Trends
P.O. Box 380
Petaluma, CA 94952
707/762-0737

International Magazines

Business International
One Dag Hammarskjold Plaza
New York, NY 10017
212/750-6300

Business Week International
McGraw-Hill, Inc.
1221 Avenue of the Americas
New York, NY 10020
212/512-2000

Columbia Journal of World Business
Columbia University
315 Uris Hall
New York, NY 10027
212/280-3431

Euromoney
Nestor House
Playhouse Yard
London, 5EX
England

Financial Times of London
14 East 60th Street
New York, NY 10022
212/752-4500

Financial Times World Business Weekly
Financial Times of London
14 East 60th Street
New York, NY 10022
212/752-4500

International Management
McGraw-Hill, Inc.
1221 Avenue of the Americas
New York, NY 10020
212/512-2000

International Trade News Letter
Norse Shipping Service
160 Broadway
New York, NY 10038

Middle East Business Intelligence
Middle East Executive Reports
717 D Street, NW, Suite 300
Washington, DC 20004
202/628-6900
(published twice a month, each issue lists existing or planned investment/trade opportunities in the Middle East, ranging from construction to manufacturing to development.)

Multinational Business
10 Rockefeller Plaza
New York, NY 10020
212/541-5730

The Economist
10 Rockefeller Plaza
New York, NY 10020
212/541-5730

World's Executive's Digest
Technology Publishing Co.
38 Commercial Wharf
Boston, MA 02110

World Marketing
Three Century Drive
Parsippany, NJ 07054
201/547-6050

APPENDIX C
How To Be An Effective Consultant on International Development Projects*

The following paper authored by Delmar Hatesohl, Assistant Director, MIAC, is based on suggestions from Dan Bigbee, Rudolph Cazenzve, Harold Dickherber, Don Esslinger, Cornelia Flor, Homer Folks, Ardyth Hanson, Roger Hanson, Geoffrey Heinrich, Jim Jorns, Wendell McKinsey, Mike Nolan, C. B. Ratchford, Jim Riordan, J. T. Scott, Darrel Watts, and L. V. Withee (Botswana ATIP Team, 1987).

▶ **NOTE** MIAC is the Midamerica International Agricultural Consortium and consists of the following member institutions:

Iowa State University
Kansas State University
University of Missouri
University of Nebraska
Oklahoma State University

Persons experienced in international development work say that consulting is a special kind of work. Those who are good at it seem to work hard at it. The successful consultant must:

1. be knowledgeable in a subject/technical field.
2. be able to apply that knowledge in a new and quite different situation.

---

* Reprinted by permission of the Midamerica International Agricultural Consortium.

3. make viable recommendations in a form and manner which increase their chances of being implemented.

The following tips on being an effective consultant come from a variety of professionals with long experience in international work. Some have done consulting, others have hired consultants, and still others have seen many consultants come and go on overseas projects.

Doing an effective job of consulting can be summarized in the following steps:

1. Understand purpose and scope of your assignment.
2. Learn all you can about the project, the country, and its people.
3. Take care of travel arrangements and logistics.
4. Start your work in-country with an open mind as to problems and solutions.
5. Turn in a written report before you leave the country, even if it is a rough draft.
6. Follow up quickly to complete all assignment details—final report and other items promised.

Following are more details under each of these steps:

Understand Purpose and Scope of Your Assignment

1. Get a clear, written set of objectives and scope for your assignment. This may take persistence on your part to get this from the contractor. You may want further clarification from the chief of party in-country.
2. Define the problem. Why has it arisen? Who are the organizations and personalities involved?
3. Who and what are behind the request for you to do this consulting? It may be host country officials or persons involved in the project or problem. It may be the advisor team in-country. Or, the request may come from the university that has a contract to fill or from the financial sponsor such as the Agency for International Development (AID), United Nations, or World Bank. Knowing who requested the help and who enthusiastically or reluctantly agreed may affect your approach.
4. Who will be your main contacts in-country? How much and from whom can you expect administrative support? Be realistic about expectations.
5. What work or action is anticipated as a result of your consultancy?
6. Know what kind of report is expected and when it is due. Prepare a brief outline for your report to avoid overlooking key areas.
7. If appropriate, ask the resident team to set up contacts for you.
8. Try to schedule your trip as nearly as possible with the time preference of the people with whom you will work.

Learn All You Can About the Project, the Country, and Its People

1. Learn about the country's history, government, trade, agriculture, customs, and culture. Beyond the usual references, other good sources are the State Department's "Post

Report" and "Area Handbook." Your university international office may have briefing material. Visit with previous consultants and students from that country.
2. If the consultancy is for an AID project, read the project paper, project agreement, previous consultant reports, annual project reports, and project evaluation reports. It may also be helpful to look at the AID congressional presentation which provides a brief summary of current and planned assistance to the country. Are there other projects operating in the country?
3. Don't assume little or nothing has been done on the subject of your consultancy. Check for sources you can ask about previous work. Have a librarian do an international search on your topic. Have other consultants been involved? It may be difficult to learn what has been done previously but it is important if you are to make the most useful recommendations.
4. Review customs and social etiquette. If you don't know the language, learn some key phrases. Learn legalities such as restrictions on camera use, etc.

Take Care of Travel Arrangements and Logistics

Work with persons designated by the contractor to help you on travel, pay, expenses, visas, and related questions. Following are points to check on:

1. Check on passport, visas, immunizations, and travel. Be sure you have the official approvals needed.
2. You may want to take an extra pair of eyeglasses or contacts, prescriptions, and other medicinal needs.
3. Carry business cards and a calculator.
4. Be prepared to provide a description of your luggage if lost.
5. Be sure you understand how and what you will be paid, what expenses will be reimbursed, and what receipts and records are needed. You may want to ask for an advance for living expenses. Determine whether your assignment will be based on a five-day or a six-day workweek.
6. Ask for enough time in-country to fully do the job you have been asked to do. If possible, arrive a day before you start a full schedule of work so that you are rested.
7. Ask about host country transportation that you will be using.
8. Prepare an itinerary with names and titles of people you will contact. Try to learn correct pronunciation of names before you arrive in-country.

Start Your Work In-Country With an Open Mind

1. Be politely aggressive and persistent in seeking information. Speak with respect and do not make unfavorable comments or comparisons about country, organizations, or facilities. Remember that you are a guest in the country; you are there to make recommendations, not to initiate action or carry it through.
2. Get both host country personnel and project team's view of the situation and alternatives. Have some standards of measurement for assessing data and other information. Be prepared to meet with people whose ideas differ strongly from your own.

3. Don't be surprised if people are reluctant to loan you documents or reports. Too many valuable papers have not been returned.
4. Listen carefully and take notes. Summarize your notes at the end of each day. Discuss them with colleagues in order to help refine your information and impressions.
5. Check your tentative report outline by the halfway mark in your consultancy. Determine how you will prepare your final report—typewriter, word processor, or other?
6. Unless you have prior approval, don't push your own research interests until you have demonstrated clearly that your first priority is to help with the local problems.
7. Allow time for social amenities. Develop friendships with host country professionals, AID personnel, and others.
8. Keep accurate record of travel itinerary, arrival and departure times, expenses, and exchange rates.

Preparing Recommendations and Your Written Report

1. Prepare a draft in time to discuss it with experienced people. This can help you avoid major errors of ideas or simple wording.
2. Remember that your recommendations must fit the country technologically, socially, and politically. Do not try to sell concepts that can't or won't be used in the local environment.
3. Your report should do the following:
   - refer to the scope of your assignment.
   - refer to previous work or reports on the same subject.
   - record factual evidence of findings. Cite sources when possible.
   - provide recommendations that are clear and actionable. Be positive in tone and tell how and why changes might be helpful.
   - avoid writing a travelogue.
4. Discuss your recommendations with the in-country person who will have responsibility for follow-up. That person may have helpful ideas or additional information.
5. Be prepared to explain why you said what you did.
6. If appropriate, look for elements of continuity and ways to build on past work you have done.
7. Leave a report with your client before you leave the country, even if it is a rough draft.
8. Be sensitive to the implications of the information you develop during your consultancy. Know who will see and be influenced by the report.

Finish up Details

Wrap up all details as quickly as possible:

1. Submit final report. A delayed report loses effectiveness.
2. Be certain you have met the terms of your contract and that whoever hired you has been so informed.

3. If appropriate, offer additional courses of action if you think your product can be expanded, reinforced, or modified.
4. Submit expense account.

Miscellaneous Tips

1. Stay out of things that are none of your business. Avoid nonrelevant advice or local personnel clashes.
2. You may be able to bring back useful documents for local library uses.
3. Remember that host country officials and team members have to deal with many consultants. Request help on only the essentials. Be prepared to live independently and sparingly.
4. Keep your sense of humor, even when you get sick, appointments with top officials fall through, and the report is due soon. Don't expect host country officials or team members to have the groundwork laid that you requested. Be flexible and ready to roll with the punches.
5. If you are on a consulting team, be sure the individual assignments are clear. Be ready to cooperate with the team leader and other team members.
6. If you plan to vacation in-country, do it after your work is finished.
7. Enjoy yourself and be appreciative of the opportunity to meet new people and situations.

Midamerica International Agricultural Consortium
Office of the Director
215 Gentry Hall
University of Missouri
Columbia, Missouri 65211 314-882-4413

APPENDIX D

Why Russia Can't Feed Itself

David Satter authored the following article, which was published in Reader's Digest, October, 1989. Permission was granted to reprint this article.

Mowcow's new spacious Kalinin Prospect supermarket was supposed to be a communist showcase. But the long lines at the counters immediately told me this was just another Soviet store.

In Russia, shoppers generally can't help themselves to food stacked on shelves. First they pay. Then they line up with their receipts at separate counters for categories like cheese, vegetables, and that most prized commodity, red meat.

On this day the store had beef, but it looked spoiled. And sausage, the staple substitute, was quickly running out.

"Sausage is finishing, sausage is finishing," announced the loudspeaker. A wave of muttering swept the meat line. Scores of shoppers, some having waited a half hour, returned for refunds before joining yet another line in the hope of buying fish or milk.

The monotony of waiting for food in Moscow is broken occasionally by the mysterious appearance of high-quality goods from abroad. Finnish butter nearly caused a stampede at another store I visited. One customer who got some for herself joked bitterly, "Thank God we didn't make Finland one of 'our' republics."

The Soviet food crisis is worse than it has been in decades, and this fall the country seems headed for yet another grain-harvest failure. Last year's shortfall, some 40 million tons, helped bring about sharply higher grain imports and shortages of everything from meat to standard consumer goods.

"For my entire life, the food situation has done nothing but deteriorate," one shopper told me. "But under Gorbachev, the decline is accelerating."

The massive failure of Soviet agriculture is much more than an inconvenience for the consumer. It threatens to derail Mikahail Gorbachev's whole economic-reform program. The reforms require people to work harder, but the workers have made it clear they won't do so if there is nothing to buy. "If we could put 80 kilos [176 pounds] of meat a year on the consumer's table," Gorbachev told a recent meeting of the Communist Party Central Committee, "all other problems would be less acute."

Why is Russia, a net exporter of food under the czars, chronically dependent on food imports? Why has Russia, with more land devoted to cereal production than any other country, become the world's largest importer of grains? And how—with roughly 25 percent of its labor force working in agriculture, compared with just three percent in the United States—can Russia average little more than half the U.S. output per acre?

Stalin's Choice

The answers can be found in the countryside, much of which looks to the visitor like a disaster area. Piles of unused silage rot in the rain. At the edge of dilapidated villages, lines of derelict, rusting tractors recall the aftermath of a World War II tank battle.

Popova is a village of one-story weathered cottages 120 miles west of Moscow, reachable only by dirt road. There old women draw water from covered wells, and drunks stagger from house to house in the middle of the working day.

Natalya Sergeyevna, a wizened pensioner, tells a familiar story about Popova's best men and women. "Most of the smart people went to America, and the smart people who stayed were killed," she said. "I remember one family who worked hard and did very well. They were deported to cut timber in Siberia."

Today's rural devastation is rooted in political decisions made in the late 1920s, when a free agricultural market was considered a threat to the Communist Party's control. Given a choice between total power and feeding the nation, Soviet dictator Joseph Stalin chose power. He exiled millions to the Arctic and Siberia, and herded the remaining farm workers into massive collective farms. Then he demanded more grain than they could produce. During 1932 and 1933, seven million Soviet farmers died of starvation.

Those who survived began to work in a completely new way—not as independent farmers but as rural slaves. The result was the death of initiative and the sundering of farmers' natural ties to the land.

For years, Soviet authorities blamed agricultural failure on the country's severe climate. Under Gorbachev, they began to blame the excesses of Stalin. Only now are some beginning to acknowledge that the problem is not Stalin but the communist system.

Farming by the Numbers

The hallmark of collectivization is central planning. "In essence, you work for a bureaucrat instead of for yourself," explains Anatoly Strelyani, a leading Soviet authority on agriculture. "You cannot show initiative or talent. This makes the farmer indifferent to the land. It's unnatural to farm in a prison."

In private agriculture, the rhythm of work is determined by sunshine, rainfall, and the requirements of the land. In the Soviet Union, what matters is the plan. Regional officials tell the chairman of each collective what to plant and how much. He, in turn, issues precise orders to his foremen, who instruct those who actually work the land. In this way, Moscow intends every operation to be directed, rigidly, like military maneuvers, without the individual farmer making a single decision.

To assure that commands are carried out, there are three million supervisors in the countryside, one for every ten farmers. Forms must be filled out to move animals, haul loads, and sow crops. In a single year, a 10,000 acre collective farm may produce half a ton of paper.

Alexei Durnov, chairman of the Iskra collective near Staritsa in central Russia, explained the situation on his farm: "At the moment, I raise flax, corn, peas, and oats. If I could make my own choices, I would raise barley, rye, and clover instead of peas and corn. Instead of sheep, I would keep beef cattle and pigs. I could increase production profits by 25 percent."

Besides forcing farmers to plant inappropriate crops, the plan dictates deadlines for sowing and harvesting. As long as the farm chairman meets them, he will not be held responsible for what happens to the harvest. Thus, if the plan says it is time to plow, the fields will be plowed even if the ground is so wet that tractors sink to their axles. If the plan says it is time to reap the crops, they will be cut, ready or not.

Harvest of Alcoholism

Not only does the system not work, but people don't work either. "Someone who tried to get the maximum return from the land would be a misfit," explained a beekeeper on a farm. "The general pattern is that of a perpetual strike."

At daybreak in the Ivanovo region northeast of Moscow, farmers assemble to receive their assignments. As the morning wears on, however, it becomes clear not everyone is working. Two persons in a group may be laboring while five are smoking. By early afternoon, the sound of tractors has all but disappeared.

To improve efficiency, the government tried pouring in money. Between 1971 and 1985, it invested nearly $1 trillion in agriculture, but got practically no increase in production.

Every place I visited, farmers were doing the absolute minimum. They fulfill the plan for sowing, for example, but ignore it for fertilization, which is harder to monitor. Instead of spreading manure, they burn it so no one will know it was never used. Chemical fertilizer is dumped into streams.

Plowmen are extremely negligent. A tractor driver, paid by the acre regardless of the harvest, cuts normal furrows about nine inches deep near the road. But as soon as he is far enough into the field so the collective farm's chairman will not check on him, he lifts his plow, races his engine and cuts furrows as little as two inches deep, with harmful consequences for the crop.

To a stranger, the most frightening scenes take place on payday. In one village in central Russia, farmers gather in the tractor shed or in the woods and begin a drinking binge. Almost everyone pitches in to buy vodka. By midday much of the village is drunk.

According to a study by the state-run Academy of Sciences, "In the overwhelming majority of Siberian villages, everyone drinks, from the chairman of the collective to the stableboy. To meet a sober male in a Siberian village after dark is like meeting a Martian."

Complete Chaos

The crops that do get harvested are subject to a bewildering array of barriers before they can reach consumers. There are only four trucks for each 1000 acres of plowed land in the Soviet countryside, far less than the number necessary to avoid delays. Once trucks are loaded, they have to brave primitive roads. Only about 20 percent of rural roads are paved. For much of the year the others are a sea of mud. As a result, milk often turns sour on the way to the creamery and is dumped in ditches, rivers, and ravines.

"Young bulls must be trucked 70 miles or more to a slaughterhouse," says Vladimir Tikhonov of the Soviet Academy of Agricultural Sciences. "But meat-packing plants work only at half their capacity because of poor equipment or too few workers. Meanwhile, the bulls stand for days chewing each other's tails from hunger."

What makes the chaos complete, however, is the absence of warehouses. Fewer than half of Soviet collectives have storage facilities. On average, Soviet farmers lose 20 to 30 percent of the harvested grain—about the amount their country imports. They also lose over a third of the fresh fruits, vegetables, and potatoes.

Prosperity and Envy

In an effort to deal with a worsening food situation, the Kremlin is offering farmers the opportunity to work tracts of land on their own. They are permitted to lease plots for 50 years or longer, pass the leases on to their descendants, and earn money based on what they produce.

Furthermore, the would-be renter has to contend with a society that has been brainwashed for decades to believe profit is profiteering and personal ambition is greed.

The story of "Volodya", a collective farmer in the Ivanovo region who tried the new system, is instructive. Volodya was allowed to rent 25 cows, a cow barn, and a tractor. But after he bought milking-machines, he discovered that the collective had neglected to turn on his electricity. He and his wife were forced to milk 25 cows three times a day by hand for two months.

Despite such difficulties, Volodya was industrious and prospered. His cows each produced over three gallons of milk a day—50 percent more than the collective's cows.

But Volodya's prosperity aroused the ire of his neighbors. They began demanding to be paid for the slightest favor. When he approached one foreman to borrow a harrow, the man said, "You're making a lot of money. Buy one yourself."

The neighbors' children insulted his children, stole their bicycles, and beat them up. When he complained, the parents told Volodya that their children had done "the right thing."

Envy is a powerful obstacle to individual initiative, and it is widespread. Says Yuri Chernichenko, a well-known agricultural journalist and advocate of reform: "Before, the renter had neighbors. Now he has enemies. The sight of their neighbor earning thousands infuriates them."

Democratization

The key to agricultural reform is to break with the collective system itself. This means allowing farmers not only to rent, but also to buy land and to create a free market in agricultural goods and supplies, so that a farmer need not depend on the whims of the local collective.

"No serious person would rent land," said a Ukrainian agricultural specialist. "But if the government offers to sell it, I'm ready."

For the moment, Soviet authorities are not willing to restore free markets or private ownership of land. One reason—a free market in agriculture could spur demands for the same in industrial goods.

But time is running out. A continuing exodus from the countryside is expected to create a shortage of some 14 million trained agricultural specialists by 1990. If urgent steps are not taken, the food crisis could become castastrophic.

Soviet economists agree that the rescue of agriculture is critical. "It could lead to the democratization of the whole system," says agricultural expert Strelyani. "One thing is certain. Without liquidation of the planning system and liberalization of the entire agricultural economy, no partial reforms can have the slightest results." Back in Moscow, in the long lines at the Kalinin Prospect supermarket, there is no shortage of support for Strelyani's views. "How long can this go on?" asks an angry shopper, after being informed that there is no more cheese.

"For as long as there's socialism," says a nearby woman who has been waiting patiently with her empty paper bags.

APPENDIX E
International Extension Programs for U.S. Citizens

G. Edward Schuh (Director, Agricultural and Rural Development, The World Bank, Washington, D.C.) presented the following speech at a Conference on the International Role of Extension; Future Directions, Kellogg Conference Center, Michigan State University, March 31–April 2, 1985. Dr. Schuh is currently Dean of the Humphrey Institute of Public Policy, University of Minnesota. This speech is reprinted by permission of Dr. Schuh.

Fifty years from now when economic historians write the modern history of the international economy, I am persuaded they will conclude that the twenty-year period we have just come through was the period of most rapid change in our modern history. By modern history, of course, I refer to our last hundred years.

These changes have wrought a completely new economic system for us. They have changed the context in which domestic policies effect our economy. They have changed the politics of the political process. And they have given each of us an enormously more complicated environment in which to make decisions about our everyday lives.

Unfortunately, the educational institutions in this nation have not kept up with our changing economic environment. Among those educational institutions I include our cooperative extension service. I think any review of either the curricula of our universities or of the programs of our extension services will provide all the evidence one needs. But if you want specific examples, just look at how agriculture is trying to deal with its problems, and also consider the response of the U.S. Senate to problems of Japan. Neither of them reflects any recognition that the world we live in today is any different than it was twenty years ago.

I would like to divide the remainder of my remarks into two parts. First, I would like to discuss the changes in the international economy and in how our country relates to it. Second, I will discuss the things people need to know to get along in the kind of world we now live in. At the end I will have some concluding comments.

Throughout my remarks I will tend to concentrate on economic issues, since that is my comparative advantage. But from time to time I will try to remind us of the political and social issues associated with the economics.

There are about four major changes in the international economy that have changed dramatically the economics of the U.S. economy and the economics of agriculture. Interestingly enough, the interactions among these four developments are quite great, with the result that the total effect is much greater than the sum of the parts.

Increased Dependence on Trade

International trade has grown faster than world GNP throughout the post-World War II period. In fact, it has grown at a faster rate in every year except three, and two of these were the last two years.

During the 1970's, there was a veritable explosion in international trade. The U.S. dependence on trade doubled in the period from 1970 to 1979. And if one extends the period back five years to 1965, our dependence on trade actually tripled. That is really an extraordinary development. Moreover, by the time we started the 1980s, the U.S. economy was as dependent on trade as was Western Europe as a whole and Japan.

An important issue, of course, is that as a country becomes increasingly dependent on trade, its economy becomes increasingly beyond the reach of domestic economic policy. The failure to recognize this obvious and well-known relationship has given rise to many mistakes in policy in this country, it has caused frustration, and it has caused no small amount of conspiracy-mindedness. Our economy doesn't respond the way it used to, and when it doesn't and we don't understand why, the tendency is to look for a conspiracy.

Recent experience with our commodity programs illustrates how the increased openness of our economy affects economic policy. For example, when the full costs of the farm program for 1983 are calculated, they will come out at about $30–35 billion. And that for a section of the economy that generated a net farm income of about $18 billion! Hence, it isn't that the government didn't do anything for agriculture. In fact, it did a lot. The point is that the forces of the international economy literally swamped the domestic programs—something that would not have happened 20 years ago.

So we find ourselves in a new situation in which we not only have to know what's going on in the rest of the international economy, but one in which the way our economy used to work isn't a valid model for how it works now or can be expected to work in the future.

Emergence of a Well-Integrated International Capital Market

This is probably one of the most dramatic developments we have had in the post-World War II period. If one goes back to the end of World War II, there was no such thing as an international capital market. There were some transfers of capital from one country to another, but these were on a government-to-government basis and we called it foreign aid.

Then, recall that in the 1960s there emerged something called the Eurodollar market, as European bankers discovered they could lend out the dollars they held at a profit. This market grew very rapidly, and eventually converted itself into a Eurocurrency market. This market also grew like "Topsy." But then in the 1970s, after OPEC jacked up petroleum prices, we began to hear about petro-dollars. They also burgeoned—huge numbers—with the result that the banks were challenged by national governments and international agencies to recycle them so as to keep the international economy from collapsing. This they did to a fault, of course.

The total amount of credit outstanding in the Eurocurrency market was estimated at about $1.7 trillion at the start of the 1980s. That is approximately commensurate with the total amount of international trade. Moreover, almost all countries use this international capital market. Hence, it constitutes a link among countries that is every bit as important as the link through trade.

What we see having happened on the international scene, then, is a shift away from the system that prevailed at the end of World War II, which was essentially a collection of relatively autonomous nation-states tied together with a little bit of trade. Today, we have a truly interdependent economy, with linkages through the international capital market every bit as important as the links through trade.

It is also important to note the flows through the international financial market now literally swamp the flows through trade. The most recent data suggest that international financial flows last year amounted to $40 trillion, which makes the $2 trillion in trade sound like a pittance. And in point of fact, it is. The international financial markets are now driving and dominating the system.

The Shift to a System of Bloc-Floating Currency Exchange Rates

Your eyes may all glaze over at the mere mention of an arcane subject like currency exchange rates. And if they do, that illustrates how much our economic world has changed. Twenty years ago we hardly knew what an exchange rate was in this country, let alone how the markets worked. Today, individual farmers do their hedging not in the grains market, but in the futures markets for foreign exchange. That is a measure of how sophisticated at least some producers have become.

The shift to a new exchange rate system took place in 1973. Prior to that date we conducted our trade under the old Bretton Woods system of fixed exchange rates. The new system we now have couldn't be more different, and it is difficult to imagine a more significant development for U.S. agriculture. Prior to these changes in the system, of course, monetary and fiscal policy had very little impact on agriculture. After the changes, however, monetary and fiscal policy now affects the agricultural sector through the trade sectors. Changes in monetary policy or in capital flows now induce realignments in the value of the currency. This affects how competitive we are internationally. More importantly, agriculture has now shifted from a situation in which it was almost completely isolated from the effects of monetary and fiscal policy, to a situation in which agriculture bears the bulk of the adjustments to changes in monetary and fiscal policy.

It should be noted that the changes are forced by changes in foreign demand. Moreover, under this new system, there is a direct link between financial markets and capital markets that did not exist before, with the financial markets being international in scope. It's the international flows of capital that are commodity markets all over the place.

Increased Monetary Instability

For reasons that are not completely understood, a great deal of monetary instability has emerged in the international system starting in about 1968. Thus, just about the time the system itself became more vulnerable to monetary disturbances, these disturbances themselves became both more frequent and more significant.

This increased instability is due in part to improper U.S. monetary and fiscal policies. But in today's world, monetary and fiscal policies in other parts of the world also play an important role. And of course, just learning how to manage the new system, with such large amounts of money sloshing around in it, has been a major challenge. But what we again see is a situation in which developments in other countries are as important to our own economy as developments which occur more narrowly within our own economy.

Let me conclude this part by noting that the events I have described above—which are major events by almost any standard you choose—have changed significantly our own economy, and especially the economic environment in which we operate. Wendell Wilkie said it 40 years ago with the title of his famous book, but it is even more of an imperative today that it was then. We are truly ONE WORLD. Events in other parts of the world, whether they have to do with weather in the Soviet Union, the monetary policies of Japan, or community programs in Brazil, have a significant effect on the U.S. economy—on its agricultural sector, on industry, or on the service sector. But what is equally important, and perhaps even more significant, is that what we do in the management of our economy has enormous implications elsewhere as well. The Mexicans put it very well: "Poor little Mexico; born so close to the United States and so far from God!" But this should be said with equal force about many other countries.

Thus we see that it is not just rhetoric to talk about the international role of extension. If extension is to be relevant to the problems that members of our society face, it has to address our society in the dimensions in which it actually exists. The kind of world we now live in is a truly interdependent economy, one in which developments in other countries are as important as developments in our own economy. To make informed and sensible decisions in today's world, we have to understand the world as it is, not as it used to be. The technological relations in communications, in transportation, and in information have changed our world forever. The only question is, "How quickly will we catch up?"

The Content of International Education Programs for U.S. Citizens

This topic deserves a great deal more attention than I will be able to give it. But I will paint with a big brush in order to get to some of the main issues. The emphasis is on what kinds of information U.S. citizens need to have to play their dual role in society—first, as a private decision-maker managing the affairs of their own life, and second, as an informed citizen choosing the people to represent them in our political process, and ultimately in the international system as a whole.

The first thing American citizens need to know is that we are now truly part of an international system, and that thinking about ourselves as a self-standing, independent economy is no longer relevant. Moreover, they need to know how the new system works. I have focused on the changes in the economy in the opening part of my remarks. But associated with the economy are various social and political systems. Knowledge about these is equally as important as knowledge about our economic system. Unfortunately, even our own political leaders and the managers of our major economic institutions do not seem to recognize the extent to which we are now part of an international system—one which no longer functions the way it used to.

U.S. citizens also need to know and understand the significance of inter-cultural or cross-cultural differences. As an international civil servant, it is easy for me to appreciate the significance of this problem. I live in a world in which words mean different things to different people—even though they are all speaking in the same language. That is a reflection of the different ethnic and national groups represented in the Bank.

This is a general problem, of course. It is an especially serious problem for a large, insular economy such as the United States. By nature we are, and have been, a parochial society, despite the large and sustained flows of migrants that come to us from all over the world. But the great feature of our culture and society is the extent to which we take these migrants and mold them into a common society. In today's world our citizens need to be sensitive to cross-cultural differences, and they need to know what some of these major differences are.

Our citizens also need to know a great deal about the international institutions that have a significant influence on their lives as well as on the lives of people in other countries. This includes the General Agreement on Tariffs—the so-called GATT, the International Monetary Fund, the World Bank, the Food and Agricultural Organization, and the UN itself. Not only do we need to understand these organizations, we need to understand U.S. policy towards them, now and in the past.

I'm always amazed at how uninformed we are. When people discuss the identity of my employer, for example, they either think I am part of a Wall Street firm, or part of the U.S. bureaucracy. For the record, the World Bank is neither. Our Secretary of Agriculture and U.S. farmers threaten to take the EC to the GATT over trade issues, not seeming to recognize that the U.S. has led the charge in insisting that the provisions of the GATT be suspended whenever there is a conflict with domestic commodity programs. In fact, we insisted that such a provision be part of the GATT from the beginning.

The U.S. and other industrialized countries starve the IMF for resources, with the result that the world is suffering a major liquidity crisis. At the same time, the U.S. fails to manage its own money supply in a fashion in keeping with its role as central banker for the world. We are on a course to disaster if we persevere in such policies. But how many of our citizens even understand the issues, let alone the facts of the situation?

Our citizens need to know something about the major forces driving the international system, and how they affect the United States and its economy. The U.S. is at the nexus of a number of major international political struggles. We are one of the centerpieces of the East-West struggle that has dominated the post-World War II period. But how many of us really understand what the issues are? For that matter, how many of our citizens have ever been exposed to Marxist thought?

The struggle of the developing countries—and the so-called North-South debate—is another major political issue on the international scene. Although we as a nation tend to put this on the back burner, it is without doubt far more important to our future than the East-West conflict, and it is difficult to imagine a world in which our citizens are more poorly informed. For example, there is nothing that will make an audience of U.S. citizens more hostile than to confront them with data showing how poorly we have done on foreign aid. Everybody "knows" we are the most benevolent society around—never mind the facts.

Similarly, few people in or outside the government seem to recognize that while we are lecturing other countries to get their economic houses in order, they can do so only if we do the same. We all face a common set of constraints. Burden sharing is not just a political slogan. It is an economic reality.

Finally, our citizens need to know as much as possible about those parts of the world affecting their vital interests. Obviously, none of us can understand everything about everything. But we can get on top of the knowledge about those parts of the world that are in our vital interests. If you are a soybean producer, for example, you need to understand how the international system works, as well as knowledge about the major consuming and producing countries. This provides an important "sorter" for identifying the detailed knowledge individuals need to acquire.

APPENDIX F
Role of the Land Grant University: An International Perspective

G. Edward Schuh (Director, Agriculture and Rural Development, The World Bank, Washington, D.C.) made the following presentation at a meeting of CARET, Washington, DC, November 7, 1987. These remarks are abridged from a paper of the same title given as University Seminar at the University of Nebraska, Lincoln, Nebraska, October 21, 1987.

▶ **NOTE** The views expressed in this paper are the author's alone and in no way should be construed as official views of the World Bank. Furthermore, Dr. Schuh is currently Dean of the Humphrey Institute of Public Policy, University of Minnesota. Dr. Schuh granted permission in order that this presentation could be reprinted.

Large continental countries tend to be insular and inward-looking. They tend to be self-contained, and to have little dependence on the rest of the world. Because they don't need the rest of the world, they tend to ignore it.

The United States is a large continental country in spades. Of special significance is the size of our economy, a fact of our existence that we still tend to ignore. In the period following World War II the United States accounted for as much as 50 percent of global GNP, with a tiny proportion of the world's population. Even today we account for approximately 30 percent—something we should keep in mind when we complain about contributing 20 percent of the budget of international organizations.

Our first President, George Washington, warned us against international entanglements. For the most part we were true to his word. It took us up until the global conflagration of World War II before we finally reached out and engaged the world on a global scale. The threat of a Soviet take-over of Europe in the immediate post-World War II period moved us to undertake the Marshall Plan, an incredible effort to assist other peoples in far-off lands. At the height of that endeavor we transferred between 3 and 4 percent of our GNP to Europe for a couple of years to help rebuild its devastated economy. Today, in contrast, we provide something like 0.1 percent of our GNP for economic assistance to other countries, and rank 17th out of the 17 industrialized countries in this share. And we're not sure we can afford that.

Even prior to the end of World War II, we actively participated in the design of an international system that has been more robust than we had a right to expect. This new system included international economic institutions such as the General Agreement on Trade and Tariffs (the GATT) and the Bretton-Wood twins—the World Bank and the International Monetary Fund. It included the United Nations system, and international forums to resolve international political difficulties. And it included a means for cultural exchange, such as UNESCO, and international development agencies such as the Food and Agriculture Organization.

At a different level, an overvalued dollar in the 1950s and 1960s enabled U.S. firms to buy up companies in other countries and thus to establish a significant presence abroad by our private sector. Parallel to that, U.S. universities and individual professionals undertook a great deal of research and teaching in other countries, in part as a component of our international development efforts.

To put it simply, the United States for a period of time overcame its isolationism to a very significant degree. But it wasn't for long. Engagement in a war in a far-off land created a crisis of confidence in our ability to engage the world in ways that were in our best interests. A dramatic fall in the value of the dollar brought us back home both economically and politically. We have dramatically reduced our foreign economic assistance. And we have turned inward once again, epitomized perhaps by the protectionist trade bill that is currently working its way through Congress.

Paradoxically, the world has become more interdependent at the very time that we are turning inward and isolationist again. Our dependence on international trade tripled from 1965 to 1979, and our economy became dramatically more open to international forces. An enormous international capital market emerged that links us to the global economy in a pervasive way, while at the same time linking our economic policies with those in other countries.

This enormous growth in global interdependence is not a happenstance. It is rooted in technological developments in the communication, transportation, and computer industries that have been without historical precedence. These technological breakthroughs have truly created a global village. Moreover, they will not go away; we will not regress technologically. Consequently, we need to turn our sights outward, in the direction in which all the economic forces are driving us.

What does all this mean for the topic of our discussion this evening? The concept of the Land Grant University emerged in a very different era, when the economic, political, and social conditions of the world were very different. We were still very much of an agrarian society at that time. A great deal of industrialization was still before us. Educational opportunities were limited, distances were longer, and communication was difficult and costly.

Many people today question whether the concept of the Land Grant University has any relevance to today's world. They look around and note that educational opportunities are abundant. They note that research is done by a wide range of institutions. And they see the private sector providing what passably provides outreach or extension services.

My perspective, of course, is that there is still a major role to be played in society by the Land Grant Universities. We still need publicly supported educational programs. We still need publicly supported research that has a strong mission orientation. And we still need extension or adult education programs.

To be successful, however, the Land Grant Universities need to redirect their programs in very significant ways, while at the same time changing the way they go about their business. The remainder of my remarks will be addressed to these issues, with the setting I have described above as the background.

Resident Instruction Programs

Most of our students will either work abroad, work for a company that has vital interests abroad, or work for a company that experiences significant competition from abroad. Alternatively, they will work in the public sector, large parts of which also are engaged with the global economy in one form or another. And most importantly, all of our students will be citizens in a society which must find its way in an increasingly competitive and complex international economy. They will need to elect political representatives who can make correct decisions in this kind of a world—decisions that will affect their well-being far into the future.

To what extent are we preparing our students for this new world into which we are emerging? My judgement is that we are doing a very inadequate job. For example, few of our students have a second language. One should contrast the number of Japanese with English-language ability with those in the United States with Japanese or Russian language ability. Language skills are the key to interacting with the rest of the world.

Our students also know precious little about the cultures in other parts of the world. Contrary to what many believe, this knowledge is not a luxury suitable only for liberal arts students, or something that is interesting but of little practical value. It is the key to understanding how people in other countries think, and how they will react to initiatives from our side, either by the private or public sector.

Our citizenry is also economically illiterate. We don't understand how our own economy operates, let alone how the economies of other countries operate. There have been dramatic changes in the international economy, as an enormous international capital market has emerged, as we have switched to a system of bloc-floating exchange rates, and as our respective economies have become increasingly open to the economic and technological forces of a global economy. Most of our citizenry do not understand even the basic principles of this new economic system. They don't understand that an economy increasingly open to international trade is increasingly beyond reach of domestic economic policy. They don't understand that with flexible exchange rates it is very difficult to dump domestic economic problems abroad. They don't understand that tariffs and other protectionist measures just shift the problem around at home, while imposing costs on the economy as a whole. Nor do they understand that protectionist measures are equivalent to an export tax, thus making us less competitive in foreign markets.

At a different level, there seems to be little recognition in this country that with flexible exchange rates, capital inflows have to be matched by a corresponding trade deficit, and that as long as we have to borrow abroad to finance our budget deficits, we will continue to incur trade deficits. Similarly, we don't seem to understand that there is a difference between foreign capital that comes in to add to our productive capacity, and that which comes in to finance our budget deficits, thus making it possible for us to sustain and increase our level of consumption.

On still another issue, it is appalling that as a citizenry we understand so little of Marxist thought—this being an important legacy of the McCarthy era. American students in our graduate programs hopelessly try to cope with dialectic thought processes on the part of their foreign student counterparts. How poorly we equip them for these interactions! And how poorly our citizenry is equipped to understand the international initiatives and dialogues of other countries—often operating from a Marxist philosophy or perspective! After all, Marxist thought tends to dominate social thought in a very large number of countries and for a large part of the world's population. Our citizens need to understand it if they are to cope with it.

Finally, our students need to know a great deal more about our own history, about the social and political systems in other countries, and about the state of development in other parts of the world. Such knowledge would not only give us greater perspective on ourselves and on our present situation, it would also enable us to engage the rest of the world more effectively.

What does all this mean for our resident instruction programs? I believe it means two things. First, we need major revisions in our curricula so that we educate our students to live and work in an international economy and society. In developing such curricula we will need to sort out what is important for students to know, and to reduce the number of courses that transmit knowledge that has a high degree of obsolescence. Second, an international perspective needs to pervade most if not all our courses and activities. It isn't so much specialized courses that we need as it is a change in perspective on the world as a whole—or on what our relevant world really is.

In a very real sense, what I am suggesting is a change in perspective from what the Land Grant Universities did in their early history. I am suggesting that we deemphasize the vocational in what we teach, and put the emphasis on the basics. In doing that, however, we are not sacrificing the basic principle of the Land Grant universities, since in a very real sense we are still preparing students for the world they will be living and working in. Moreover, the more vocationally oriented material is available from a wide number of other institutions in our society.

Research

If we are going to compete in a global economy, either economically or politically, we have to understand it. And although we engage in more research on other countries than do most other countries, what we do is still not commensurate with our continued global importance, with our global political interests, or with the benefits that can accrue to us from being more competitive in the international economy.

Many of the things I discussed in my remarks on education are pertinent to this topic. The immediate question is whether we have an adequate knowledge base to give our students the kind of education they need and deserve. I believe the answer to that question is a flat "no." The same applies to the knowledge we can provide to the private sector as it attempts to compete in the international economy.

Let me cite a few examples. The international economy has undergone a veritable transformation these past twenty years with the emergence of the international capital market, the shift to bloc-floating exchange rates, and the growth in trade. These developments have changed the way that monetary and fiscal policy effect our economy. They have changed the way that trade policy and domestic economic policies effect our economy. And they have changed the factors which determine how competitive we are in foreign markets. Yet we understand these developments very poorly. Similarly, we have not done the design work on the new institutional arrangements needed for this changed international economy. And we understand our competitors very poorly.

Let me cite just a few examples from agriculture. The United States has only a few major competitors in agricultural trade. These are the European Community, Canada, Brazil, Argentina, and Australia. Yet our knowledge base on the agriculture of those economies is very weak. Similarly, we have a very limited knowledge base on how policies in other countries effect our own markets and our competitive potential.

More generally, there are dramatic shifts in comparative advantage taking place in the international economy. A half-dozen newly industrializing countries, the so-called NIC's, located in Asia and Latin America, have pulled themselves up by their economic bootstraps by means of export-oriented industrialization drives. At the same time, international agricultural research systems are producing a new, more highly productive technology for food crops in the tropics. These new technologies promise to be the basis for more rapid economic growth in those countries. They also promise to change the basis for global comparative advantage. The research we as a nation are doing to better understand our competitive potential is precious little. Yet that surely should be high on the agenda of universities that have a mandate to promote the growth of their state and nation.

Finally, I am struck by how little we do to keep up with the new knowledge that is being produced in other countries. There was a time when the United States had a virtual monopoly on the research being done on the frontier of knowledge, and was at the same time the technological leader of the world. Unfortunately, we believed our own rhetoric and became complacent. Other countries have geared up their scientific and technological institutions and have either closed the gap with us, or in some cases actually taken the lead.

Clearly, some part of that shift in balance was inevitable, and it would take a herculean effort for the United States to stay out in front in everything. What is troublesome, however, is how little we have done to access the knowledge that is being produced in other countries. There are a lot of things we could do in this regard. But research to discover what research is being done elsewhere would be a very useful starting point.

Extension or Adult Education

The rapid changes in the international economy leave our citizens with a knowledge gap that is ever greater. Our educational systems did little to prepare them for the kind of world in which they now must function. The means by which they can catch up are sorely limited and that reflects badly on our extension systems.

Many in this country argue that our agricultural extension system is a relic of the past and no longer needed. I would argue just the contrary. Given the rapid rate of change in our economy, driven in large part by developments in other parts of the world, the need for an effective extension or adult education systems has probably never been greater, nor the potential payoff from it ever higher.

But here, too, the mission needs to be changed. It's true that the private sector can provide a great deal of what passes for technical assistance, just as it does in providing the professional services of medicine and law. But extension and adult education now needs to focus on a different set of issues, and to broaden its base to include inputs from across the university.

My concept of a modern extension system is one that essentially adapts regular resident instruction to the special needs of clientele groups and offers it at convenient times and places for adult students. The courses should address most of the same subject matter I outlined under education above, and it may make sense to offer formal credit for the courses. In effect, we should develop the concept of continuous education, or continuous learning.

In addition to this formal training, our extension services need to develop more effective public affairs programs. The need in the international arena is indeed great. We simply will not obtain more rational economic policies until our citizens have an adequate knowledge base to vote intelligently on the issues and in their choice of political representation. Our citizens are sorely deficient in their knowledge of international institutional arrangements such as the GATT, the International Monetary Fund, the World Bank, the OECD, the U.N., and so on. Yet these are the very institutions that have taken on increased importance in our lives.

International Burden Sharing

In drawing these comments to a close, I would like to make some remarks on international burden sharing. As I noted earlier, we have a tendency in contemporary America to think that we can ignore the rest of the world, and that the international debt problem and the problems of poverty are for somebody else to resolve. The question is often raised, "Why are we giving money to those foreigners when we have so many problems here at home?"

That plaintive plea fails to recognize the extent to which we truly are part of an interdependent international economy. The point is that spending our money abroad may be the most effective way to help ourselves here at home. Whether we like it or not, we share in these international problems. The only question is in what form we will share them. Sluggish economic growth in developing countries means sluggish economic growth for the United States. A failure to service their international debts on the part of the developing countries will put our financial system at risk.

A serious concern about the developing countries is not something we should do just out of a sense of philanthropy. We should have a concern because our fate is intricately intertwined with theirs. U.S. commodity groups simply have it wrong when they argue that agricultural development in low income countries is not in their best interests. This is a classic case of our producers shooting themselves in the foot.

The same applies to developments in the other industrialized countries and in the centrally planned economies. The Common Agricultural Policy of Western Europe has significant impacts on economic prosperity in the United States. We need to understand that policy. Similarly, significant liberalization in the Soviet Union might open up enormous trade opportunities for the United States, just as liberalization in China did. Will we be prepared for it? More importantly, do we know how to bring about that liberalization?

Concluding Comments

One of the things that struck me in preparing these comments was the extent to which the need to adjust to and adapt to a global economy in which distances and times are shortened and in which competition from abroad has become real has made the Land Grant concept more relevant to the contemporary problems of our day rather than less relevant. The competitive challenge from abroad has increased dramatically the demand for new knowledge and the demand for new skills on the part of our students. Just as the need to remain internationally competitive is causing us to rethink and revitalize our primary and secondary educational systems, it should also cause us to rethink the mission and mandate of our Land Grant Universities. Those of us on the faculties of these Universities have major challenges before us if we are to revitalize them and re-establish their relevance to the many problems our society faces. Similarly, those who support these universities have a responsibility to see that they are moving in the right direction. We all need to work together to assure that we are truly meeting the needs of our society—which after all, is what the Land Grant concept was all about!

APPENDIX G
Basic Principles for College and University Involvement in International Activities

This document was prepared by the International Affairs Committee of the National Association of State Universities and Land Grant Colleges. Endorsed by the Association Executive Committee May 17, 1983. Endorsed by the Association of U.S. University Directors of International Agricultural Programs (AUSUDIAP) June 23, 1983. (Reprinted with Permission.)

Introduction

In 1979, the National Association of State Universities and Land Grant Colleges adopted a "Statement of Principles for Effective Participation in International Development Activities." The main purposes of the Statement are to provide guidance for maintaining and improving professional practices in international development activities to institutions already involved in such activities; and, to assist those institutions who are planning to be involved in international development activities as they prepare themselves for active participation. The present document is an elaboration of the original Statement and provides more detailed guidelines for internal self-study.

For the purpose of this document, the term "International Development Activities," considered in broad perspective, refers not only to contracting for and operation of technical assistance projects abroad, but also to the integration of international development activities into appropriate and relevant on-campus programs.

The Basic Principles

Principle 1. Effective participation in international development activities requires a commitment by both administration and faculty.

Commitment, in this sense, means a deliberate and considered intent, plan, and effort to include international development activities as an integral part of the institution's ongoing programs. It means an intent to give administrative and policy support to development activities to the extent necessary to accomplish the same quality of performance as in domestic activities. This commitment should be evidenced, as appropriate, at each level in the university—the governing board, the central administration, the college, the department, and the individual faculty member.

Commitment implies a belief in the inherent value and importance of such activities, that they are worthy of scholarly endeavor, and that U.S. universities have unique capacities to contribute to U.S. and worldwide goals in development.

Such intents and beliefs are essential to the establishment of priorities, the assignments of resources, and the generation of sufficient determination to achieve the maximum performance on contracts, training programs, and other development activities undertaken by the university.

Principle 2. Effective involvement in international development activities should be consistent with the institution's mission, commitment, and competencies.

Since human resources available to any institution are limited, no institution can address adequately the full gamut of opportunities for involvement. Resources committed to activities that are contributory to the basic program(s) of the institution will be more productive and will yield more meaningful feedback. Activities consistent with the basic mission are more likely to be understood and supported, both by the university and by its constituency.

Development activities demand a high degree of capability from well qualified professionals performing in a complex environment. Problems of development are exceedingly complex and their resolution requires the best professional effort. Optimum results are possible only if competent personnel in adequate numbers are assigned to international development activities undertaken.

Principle 3. Requisite key and supporting personnel resources must be available to assure effective, responsible, and continuous involvement in each project undertaken.

The personnel in a development activity constitute the key to its success. Requisite personnel resources embody an adequate number of persons with specific qualifications for the task at hand.

In considering development activity opportunities, the university should recognize the advantage of obtaining the requisite personnel resources from its own faculty for any project it chooses to undertake. Nevertheless, the extent to which every project can be staffed fully from the institution's own faculty is limited.

The identification of project leadership, staffing of a majority of the key positions, and provision for adequate campus backstopping by appropriate qualified faculty are minimums for which to strive.

The assignment of well qualified professionals, from whatever source, for development positions is the primary goal. Collaboration of two or more universities through joint arrangements facilitates achievement of personnel goals appropriate to this principle.

Principle 4. Adequate incentives should exist to assure that high-quality, professionally active faculty members become involved in development activities.

Incentives to faculty members include professional recognition, professional advancement, opportunities for professional growth, pursuit of research goals, salary increments directly associated with overseas assignments, and salary levels tied to the professional qualifications of each faculty member. Incentives in this context also include encouragement to the department to participate in the international activity and to support the participation of particular faculty members. Such incentives should be codified in personnel policies and practices and communicated to the faculty and to departmental administrators.

These incentives are justified in that they are necessary to attract the professional expertise required in international activities. Well qualified, senior faculty can be attracted to these activities and away from other challenging opportunities within the institution only through meaningful incentives. Faculty members with identified interest in international activities should receive as much incentive to work within the university as outside.

Principle 5. Adequate and timely logistical support of, and professional service to, a faculty member or a team abroad requires special administrative policies and practices.

Logistical support includes such things as placing and managing participants in training programs, planning and managing predeparture activities for overseas personnel, purchase and shipment of commodities, and the management of personnel concerns such as salary, per diem, and records.

Professional services include backup and support of colleagues in whatever endeavor the overseas faculty member undertakes, and any other activity that contributes to professional growth while on duty abroad.

The extent to which an institution needs to make special provision for these services depends upon the level of involvement. Adequate provision is less expensive if an institution supplies only one or more faculty members in contrast to one that assumes leadership, either as a contractor or a lead in a multi-university project. Some universities prefer to supply faculty for projects in which another university has primary leadership and management responsibility. In such joint activities, the responsibility of each institution should be clearly defined.

By the nature of development activities, professionals on international assignment are somewhat isolated professionally, at least in comparison to their colleagues at home. A continuous flow of up-to-date information about their profession and about their university improves productivity, both through knowledge that there is a mechanism for keeping up to date, and through the usefulness of new information in the exercise of the job assignment. To provide this function well, the backup professional should be as knowledgeable of the international assignment as is the colleague on assignment.

Logistical support performed in the United States is beyond the control of the individuals abroad, yet impinges significantly on their performance level. It also has an effect on the image of the contracting institution that is held by the funding agency and the host country personnel and agencies. Such support should be anticipated and appropriate arrangements made to provide it.

Principle 6. Provision of adequate orientation and specialized training of project personnel is necessary, especially before departure for international assignments.

The object of this orientation and training should be to develop an understanding and appreciation of the setting in which the project is to be undertaken.

The nature of social, political, and economic institutions and practices in developing countries is related to the culture of that country. The success of efforts to introduce innovation into these institutions or related to these practices is dependent to a considerable extent upon understanding the culture in which they operate. Language capability is especially important in the transmission of ideas and in the comprehension of institutions and practices of the country. The impact of such understanding and capability on the level of performance is obvious.

Principle 7. Teaching, research, and public service activities of the university are enhanced by properly selected and executed international development activities, followed by appropriate integration efforts.

The fundamental reason for university involvement in international activities is that they relate and contribute to the basic mission of the university. The integration of an international dimension as reflected by the presence of international students and scholars and of faculty with overseas experience is important to provide a rounded and high-quality educational experience to all students, and in the achievement of high-quality research and extension programs. Establishing continuous relationships with counterpart institutions abroad is an important way to enhance fulfillment of the universal and worldwide mission of the university community, and is a desirable facet of any international activity undertaken.

The selection of projects that will contribute to the enhancement noted above is likely directly associated with superior performance in the execution of that project.

The extent to which experiences in previous international development activities are integrated into university programs and shared with other faculty will affect the capacity of the university to respond to future opportunities and the effectiveness of the university in execution of future activities.

Principle 8. Adequate and appropriate training for international students, particularly through contract training programs, depends on specially focused university policies and practices to deal with the students' unique needs and background, and the highly specialized requirements of the training program.

International students bring with them a varying mixture of cultural, personal, economic, and language characteristics. They come with different types of educational experiences. Their expectations may be substantially different from those of U.S. students. They may have little understanding of the student role as defined in the United States. They encounter special requirements related to matriculation, advising, programming, and housing. Their unique needs should be recognized by host universities, and procedures identified to deal with them in a systematic fashion.

Principle 9. Internal evaluation procedures are necessary to provide for continuous monitoring of activities, including international, and prompt adjustments when needed for international development activities.

In the execution of international development activities, a university faces a greater than normal risk in the possible mismatch of personnel with project goals, in error in planning, and from unpredictable interruption and delay of project activity. The context in which an activity operates in a developing country changes over time. In addition, project progress may be less than adequate under these conditions, and replacement or augmentation of project inputs identified in the original project design may be necessary. Personnel may need to be replaced.

For all these reasons, a specially designed procedure is necessary to monitor continuously the problems encountered, the accomplishments achieved, new elements in the project environment, and to adjust objectives and procedures to maximize progress toward established goals.

References

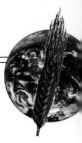

CHAPTER 1

Anderson, D. Craig. "Agricultural Education: Definitions and Implications for International Development." NACTA JOURNAL, June 1984.

Hamilton, John Maxwell. MAIN STREET AMERICA AND THE THIRD WORLD. Cabin John, Maryland: Seven Lockes Press, 1986.

Meaders, O. Donald, et al. AGRICULTURAL EDUCATION IN AFRICA: A SYSTEM VIEW. Michigan State University, Agricultural and Extension Education and Institute of International Agriculture, September 1988.

Schuh, G. Edward. "International Extension Programs for U. S. Citizens." Presentation made at the Conference on the International Role of Extension: Future Directions, Kellogg Conference Center, Michigan State University, March 1985.

Schuh, G. Edward. "Role of the Land Grant University: An International Perspective." Presentation made at the University Seminar at the University of Nebraska, October 1987.

"Sharing the Global Harvest: A Kinship of Farmers." New York: CARE, Inc., 1991.

Swanson, Burton E. AGRICULTURAL EXTENSION: A REFERENCE MANUAL. Second Edition. Rome, Italy: Food and Agriculture Organization of the United Nations, 1984.

Yeutter, Clayton. Address by Secretary of Agriculture Clayton Yeutter to the United Nations. New York, October 1989.

CHAPTER 2

THE CHALLENGE OF WORLD HUNGER (Study Guide). New York: Bread for the World Educational Fund, 1980.

McCleary, Paul and J. Philip Wogaman. QUALITY OF LIFE IN A GLOBAL SOCIETY. New York: Friendship Press, 1978.
OUR WORLD. Washington, D.C.: National Geographic Society, 1990.
"Population and Progress: The Facts." NEW INTERNATIONALIST, No. 79, September 1979.
Presidential Commission on World Hunger Preliminary Report. Washington, D.C.: December 1979.
Wennergren, E. Boyd and William Furlong. THE UNITED STATES AND WORLD AGRICULTURAL DEVELOPMENT. National Association of State Colleges and Land Grant Universities. Washington, D.C.: 1985.
"World Agriculture: Review and Prospects into the 1990s." Winrock International. Morrilton, Arkansas: 1983.
Wortman, Sterling and Ralph W. Cummings. TO FEED THIS WORLD. Baltimore: Johns Hopkins University Press, 1978.

CHAPTER 3

Avery, Dennis T. GLOBAL FOOD PROGRESS. Hudson Institute, Indianapolis: May 1991.
Bergland, Bob. "Attacking the Problem of World Hunger." NATIONAL FORUM, LXIX, No. 2. Phi Kappa Phi. Baton Rouge, Louisiana: 1979.
McCleary, Paul and J. Philip Wogaman. QUALITY OF LIFE IN A GLOBAL SOCIETY. New York: Friendship Press, 1978.
McPherson, M. Peter. "World Food Day." AGENDA, 4, No. 8. Washington, D.C.: Agency for International Development, 1981.
Minear, Larry. NEW HOPE FOR THE HUNGRY. New York: Friendship Press, 1975.
Persons, Edgar. "The Visitor." Department of Agricultural Education, University of Minnesota, Vol. LXXVII, No. 3. St. Paul, Minnesota: Summer 1991.
PRESIDENTIAL COMMISSION ON WORLD HUNGER REPORT (abridged version). Washington, D.C.: U.S. Government Printing Office, 1980.
Thomas, G. W., S.E. Curl, and W. F. Bennett, Sr. FOOD AND FIBER FOR A CHANGING WORLD. Danville, Illinois: Interstate Publishers, 1979.
Wennergren, E. Boyd and William Furlong. THE UNITED STATES AND WORLD AGRICULTURAL DEVELOPMENT. National Association of State Colleges and Land Grant Universities. Washington, D.C.: 1985.
Williams, Sue Rodwell. NUTRITION AND DIET THERAPY. St. Louis, Missouri: C. V. Mosby Co., 1969.
World Food Conference Proceedings. Rome, Italy: United Nations World Food Organization, 1974.

CHAPTER 4

Anderson, James R. A GEOGRAPHY OF AGRICULTURE. Dubuque, Iowa: William C. Brown Company Publishers, 1970.
Brochure. Communicating for Agriculture Exchange Program. Ames, Iowa: 1991.
Coffin, Tristram, ed. "Food and Energy: Promising New Leads." THE WASHINGTON SPECTATOR, 4, No. 13, 1978.

David, Steven R. "The Third World and the Soviet Union." WORLD DEVELOPMENT FORUM, Vol. 2, No. 14, August 1984.

Duckham, A. N. and G. B. Masefield. FARMING SYSTEMS OF THE WORLD. New York: Praeger Publishers, 1969.

FRONTLINES. The U.S. Agency for International Development. Washington, D.C.: July 1991.

Hodgson, Harlow J. "Forages, Ruminant Livestock and Food." BIOSCIENCE, 26, No. 10, October 1979.

Thomas, G. W., S. E. Curl, and W. F. Bennett, Sr. FOOD AND FIBER FOR A CHANGING WORLD. Danville, Illinois: Interstate Publishers, 1976.

Kendrick, J. B. "The Other Energy Crisis." Seminar Report. PROTEIN NUTRITION IN THE WORLD'S FOOD SUPPLY. Davis, California: University of California, April 1976.

Kellog, Charles E. AGRICULTURAL DEVELOPMENT, SOIL, FOOD, PEOPLE, WORK. Madison: Soil Science Society of America, Inc., 1975.

Lofchie, Michael F. THE POLICY FACTOR: AGRICULTURAL PERFORMANCE IN KENYA AND TANZANIA. Boulder, Colorado: Lynne Rienner Publishers, 1988.

Muhlfield, Liz. "Fish, the Need to Know." RF ILLUSTRATED 3, No. 2, Washington, D.C. December 1976.

OUR WORLD. Washington, D.C.: National Geographic Society, 1990.

Spitzer, Robert R. NO NEED FOR HUNGER. Danville, Illinois: Interstate Publishers, 1981.

"World Agriculture: Review and Prospects into the 1990s—A Summary." Winrock International. Morrilton, Arkansas: 1983.

Wennergren, E. Boyd and William Furlong. THE UNITED STATES AND WORLD AGRICULTURAL DEVELOPMENT. National Association of State Colleges and Land Grant Universities. Washington, D.C.: 1985.

Wolf, Anthony. "Are We Heading for the Ninth Ice Age?" RF ILLUSTRATED, 2, No. 4, Washington, D.C. March 1976.

Weisner, Robert and Massoud Denbaly. "Cassava Root Production." WORLD FOOD TRADE AND U.S. AGRICULTURE 1960–1981. Ames, Iowa: World Food Institute, 1982.

Wortman, Sterling and Ralph W. Cummings. TO FEED THIS WORLD. Baltimore: Johns Hopkins University Press, 1978.

CHAPTER 5

Brown, Lester. "Vanishing Croplands." AGENDA, 1, No. 11. Agency for International Development. Washington, D.C.: December 1978.

DEVELOPMENT FORUM. Geneva, Switzerland: United Nations University and UN Division of Economic and Social Information. May 1982.

Edrst, Jeff. "The Developmental-Environmental Connection." AGENDA. Agency for International Development. Washington, D.C.: May 1981.

Furtick, William R. "International Concern for the Environment Brings New USAID Initiatives." INTERNATIONAL AGRICULTURE, Vol. 4, No. 4. Champaign, Illinois: University of Illinois at Urbana-Champaign, 1989.

Gallon, Gary. "Assessing the Impact." DEVELOPMENT FORUM, X, No. 4. Geneva, Switzerland: United Nations University and UN Division of Economic and Social Information, May 1982.

Spitzer, Robert R. NO NEED FOR HUNGER. Danville, Illinois: Interstate Publishers, 1981.

Wolff, Anthony. "Fragile Lands." RF ILLUSTRATED. New York: The Rockefeller Foundation, September 1978.
York. E.T. "Sustainable Agriculture Production." INTERNATIONAL AGRICULTURE, Vol. 4, No. 4. Champaign, Illinois: University of Illinois at Urbana-Champaign, 1989.

CHAPTER 6

Ensminger, Douglas. "Constraints to Millions of Small Farmers in Developing Countries." WORLD FOOD CONFERENCE PROCEEDINGS. Ames, Iowa: Iowa State University, 1976.
Food and Energy Council. FUTURE WORLD: FOOD AND ENERGY. Columbia, Missouri: 1977.
Kendrick, J.B. "The Other Energy Crisis." Seminar Report. PROTEIN NUTRITION IN THE WORLD'S FOOD SUPPLY. Davis, California: University of California, April 1976.
Newendorp, Paul D. ENERGY—WHAT WILL OUR PRIORITIES BE? Study Paper. Oklahoma Cooperative Schools of Christian Mission. Oklahoma City, Oklahoma: The United Methodist Church, 1982.
McPherson, M. Peter. "Continuing a Proud Tradition of Helping Others." AGENDA. Washington, D.C.: Agency for International Development, May 1981.
Obeng, Henry B. "Natural Resources Currently Available for Crop and Animal Production." WORLD CONFERENCE PROCEEDINGS. Ames, Iowa: Iowa State University, 1976.
Schurr, Sam H. ENERGY, ECONOMIC GROWTH, AND THE ENVIRONMENT. Baltimore: Johns Hopkins University Press, 1972.
Steinhart, John S. and Carol E. Steinhart. "Energy Use in the U.S. Food System." SCIENCE 184, April 1974.
The Lik Lik Bok, Manual of Appropriate Technology, New Guinea Adult Extension Department, Papua New Guinea, 1977.
Thomas, G. W., S. E. Curl, and W. F. Bennett, Sr. FOOD AND FIBER FOR A CHANGING WORLD. Danville, Illinois: Interstate Publishers, 1976.

CHAPTER 7

Aritaratne, A.T. "Pitfalls and Promises: Choosing the Right Technological Road." DIALOGUE. Washington, D.C.: International Volunteer Services, Winter/Spring 1977/1978.
Brochure. Winrock International. Morrilton, Arkansas: 1991.
Brochure. HEART Program, Warner Southern College. Lake Wales, Florida: 1991.
Brochure. Heifer Project International. Little Rock, Arkansas: 1991.
Christiansen, James. Personal Communique. Texas A & M University. College Station, Texas, 1992.
Finley, Eddy. Personal Tour of Heifer Project International and the International Livestock Learning Center. Perryville, Arkansas, March 1987.
Finley, Eddy. Personal Tour of the HEART Program. Lake Wales, Florida, March 1988.
Finley, Eddy. Personal Tour of WINROCK INTERNATIONAL. Petit Jean Mountain, Morrilton, Arkansas, March 1987.
Goulet, Dennis A. THE CRUEL CHOICE. New York: Atheneum, 1971.
Jedlicka, Allen D. ORGANIZATION FOR RURAL DEVELOPMENT: RISK TAKING AND APPROPRIATE TECHNOLOGY. New York: Praeger Publishers, 1977.

Norman, Colin. "Mass Production or Production by the Masses" (as excerpted from Worldwatch Paper #21). AGENDA, 1, No. 7. Washington, D.C.: Agency for International Development, 1978.

Sanchez, Vincente and Fernando Monasterio. "How to Manage Nature." DEVELOPMENT FORUM, VIII, No. 1. Geneva, Switzerland: United Nations January/February 1980.

Spitzer, Robert R. NO NEED FOR HUNGER. Danville, Illinois: Interstate Publishers, 1981.

WORLD DEVELOPMENT FORUM. Vol. 2, No. 14. Washington, D.C.: August 1984.

Wortman, Sterling and Ralph W. Cummings. TO FEED THIS WORLD. Baltimore: Johns Hopkins University Press, 1978.

CHAPTER 8

Clausen, A.W. Mimeo paper, "Remarks before the United Nations General Assembly," November 12, 1982.

Ebeling, Walter. THE FRUITED PLAIN. Berkeley, California: University of California Press, 1979.

FARM MANAGEMENT: HOW TO ACHIEVE YOUR FARM BUSINESS GOALS. USDA Yearbook of Agriculture. Washington, D.C.: U.S. Government Printing Office, 1989.

"Aspects of the World Feed-Livestock Economy." Food and Agriculture Organization of the United Nations. Economic and Social Development Paper No. 57. Rome, Italy: 1986.

"The World Banana Economy." Food and Agriculture Organization of the United Nations. Economic and Social Development Paper No. 76. Rome, Italy: 1988.

"World Market for Horticulture Products." Food and Agriculture Organization of the United Nations. Economic and Social Development Paper No. 80. Rome, Italy: 1989.

"The State of Food and Agriculture." Food and Agriculture Organization of the United Nations. Rome, Italy: 1990.

Foreign Agricultural Service. Foreign Agriculture, Vol. 26, No. 2. Washington, D.C.: U.S. Department of Agriculture, February 1988.

Halcrow, Harold A. FOOD POLICY FOR AMERICA. New York: McGraw-Hill, 1977.

Kodatchenko, A. "Air-trade Links." DEVELOPMENT FORUM 8, No. 7. Geneva, Switzerland: United Nations University, DES1/DPI. September 1980.

Manfredi, E.M. and D.O. Flynn. "Nations Trade Food for Everyone's Benefit." USDA Yearbook of Agriculture. Washington, D.C.: U.S. Government Printing Office, 1981.

Martin, Torence. "Trade, A Hunger Issue." BREAD FOR THE WORLD BACKGROUND PAPER #31. February 1979.

Morgan, Dan. MERCHANTS OF GRAIN. New York: Viking Press, 1979.

Robinson, Kenneth L., Bob F. Jones, and Lois A. Simonds. FOOD POLICY BASEBOOK, WHO WILL GET IT? Ohio State University, National Policy on Education Committee. Columbus: 1975.

Spitzer, Robert R. NO NEED FOR HUNGER. Danville, Illinois: Interstate Publishers, 1981.

Toton, Suzanne C. WORLD HUNGER, THE RESPONSIBILITY OF CHRISTIAN EDUCATION. Maryknoll, New York: Orbis Books, 1982.

Thomas, G. W., S. E. Curl, and W. F. Bennett, Sr. FOOD AND FIBER FOR A CHANGING WORLD. Rev. Ed. Danville, Illinois: Interstate Publishers, 1982.

United States Department of Agriculture. ECONOMIC RESEARCH SERVICE REPORT #98. Washington, D.C.: U.S. Government Printing Office, 1974.

Wennergren, E. Boyd and William Furlong. THE UNITED STATES AND WORLD AGRICULTURAL DEVELOPMENT. National Association of State Colleges and Land Grant Universities. Washington, D.C.: 1985.

World Food Institute. WORLD FOOD TRADE AND U.S. AGRICULTURE. Ames, Iowa: Iowa State University, 1981.

CHAPTER 9

Crosson, Pierre R. "Institutional Obstacles to Expansion of World Food Production." SCIENCE, 1988, No. 4188. May 9, 1975.

Haig, Alexander M., Jr. "Development Assistance: International Challenge." AGENDA, 4, No. 4. Washington, D.C.: Agency for International Development, 1978.

Howe, James W. THE U.S. AND WORLD DEVELOPMENT: AGENDA FOR ACTION. New York: Praeger Publishers, 1974.

Laidlow, Karen A. "Food Crisis: A Political Matter?" AGENDA, 1, No. 9, Washington, D.C.: Agency for International Development, 1978.

Oluwasanmi, H.A. "Socio-Economic Aspects of Feeding People." WORLD FOOD CONFERENCE OF 1976. Ames, Iowa: Iowa State University, 1976.

Penn, J.B. "Poland's Agriculture: It Could Make the Critical Difference." CHOICES. Fourth Quarter. Herndon, Virginia: 1989.

"Resources for the Future Agricultural Policies in a New Decade." National Planning Association. Washington, D.C.: 1991.

"USAID Fights Starvation in Africa." USAID HIGHLIGHTS, Vol. 7, No. 3. Washington, D.C.: U.S. Agency for International Development, Summer 1991.

Walters, Harry. "Difficult Issues Underlying Food Problems." SCIENCE, 188, No. 4188. May 9, 1975.

Wennergren, E. Boyd and William Furlong. THE UNITED STATES AND WORLD AGRICULTURAL DEVELOPMENT. National Association of State Colleges and Land Grant Universities. Washington, D.C.: 1985.

Wharton, Clifton, Jr. "Are Politicians the Missing Link?" RF ILLUSTRATED, 3, No. 4. September 1977.

CHAPTER 10

"Agency Rushes to Kurdish Refugees." USAID HIGHLIGHTS, Vol. 7, No. 3. Washington, D.C.: U.S. Agency for International Development, Summer 1991.

A GUIDE TO TITLE XII AND BIFAD (Booklet). Washington, D.C.: Board for International Food and Agricultural Development/Agency for International Development, November 1983.

Bloch, Julia Chang. "Food for Peace: Where is it Going?" AGENDA, 4, No. 8. Washington, D.C.: Agency for International Development, 1981.

Bolling, Landrum R. and Craig Smith. PRIVATE FOREIGN AID. Boulder, Colorado: Westview Press, 1982.

Chernush, Kay. "Growing Together, Peace Corps and AID." AGENDA, 4, No. 8, Washington, D.C.: Agency for International Development, 1981.

Day, Terrence L. "Universities go International." AGENDA, 4, No. 9, Washington, D.C.: Agency for International Development, 1981.

Halcrow, Harold G. FOOD POLICY FOR AMERICA. New York: McGraw-Hill, 1977.

"Helping People Through Agriculture" (Brochure). MIAC MidAmerica International Agricultural Consortium. Lincoln, Nebraska: 1990.

UNITED NATIONS BASIC TEXTS. Vols. I and II. Rome, Italy: FAO, 1974 and 1979.

CHAPTER 11

Oppenheim, Peter K. INTERNATIONAL BANKING. The American Bankers Association, American Institute of Banking, 1970.

Price, Robert R. "Agricultural Education Programs in Higher Education." Report of Tour and Study in Saudi Arabia; International Programs, Oklahoma State University, Stillwater, Oklahoma, May 1980.

UNITED NATIONS DEVELOPMENT PROGRAMME (Brochure). Geneva, Switzerland: United Nations World Food Organization, 1981.

Wennergren, E. Boyd and William Furlong. THE UNITED STATES AND WORLD AGRICULTURAL DEVELOPMENT. National Association of State Colleges and Land Grant Universities. Washington, D.C.: 1985.

WORKING PAPER: A FOOD POLICY AGENDA FOR THE 1980s. Washington, D.C.: Interreligious Task Force on U.S. Food Policy, January 1980.

THE WORLD BANK ANNUAL REPORT, 1982. Washington, D.C.: The World Bank, 1983.

WORLD BANK NEWS, II, Nos. 21, 22, 23. Washington, D.C.: The World Bank, 1983.

Wortman, Sterling and Ralph W. Cummings. TO FEED THIS WORLD. Baltimore: Johns Hopkins University Press, 1978.

CHAPTER 12

Bunch, Ronald. TWO EARS OF CORN: A GUIDE TO PEOPLE CENTERED AGRICULTURAL IMPROVEMENT. Oklahoma City, Oklahoma: World Neighbors, 1982.

Donaldson, Les and Edward E. Scannell. HUMAN RESOURCE DEVELOPMENT—THE NEW TRAINER'S GUIDE (Tenth Printing). Reading, Massachusetts: Addison-Wesley Publishing Company, 1986.

Jedlicka, Allan D. ORGANIZATION FOR RURAL DEVELOPMENT. Westport, Connecticut: Greenwood Publishing Group, 1977.

Knowles, Malcolm S. THE MODERN PRACTICES OF ADULT EDUCATION. New York: Associated Press, 1971.

Lionberger, Herbert F. and Paul H. Guin. COMMUNICATION STRATEGIES: A GUIDE FOR AGRICULTURAL CHANGE-AGENTS. Danville, Illinois: Interstate Publishers, 1982.

Okatahi, Stephen S. and Richard F. Welton. "NigerianStudy Reports Teacher Professional Competencies Needed by Agricultural Colleges." NACTA JOURNAL, Vol. XXIX, No. 1, March 1985.

Wennergren, E. Boyd and William Furlong. THE UNITED STATES AND WORLD AGRICULTURAL DEVELOPMENT. National Association of State Colleges and Land Grant Universities. Washington, D.C.: 1985.

Reisbeck, Robert F. "Teaching/Learning." Communications Training Bulletin, No. 7, Oklahoma State University, Cooperative Extension Service. Stillwater, Oklahoma: 1976.

WORLD CONFERENCE ON AGRICULTURAL EDUCATION AND TRAINING REPORT, Vol. 11:67. FAO/UNESCO/ILO, 1970.

CHAPTER 13

Charles, Pat. "Projects in the Caribbean." INTERNATIONAL COMMUNITY EDUCATION NEWSLETTER, 5, No. 1, 1983.

Cisse, Samba Yacine. EDUCATION IN AFRICA IN THE LIGHT OF THE HARARE CONFERENCE (1982). Paris, France: UNESCO, 1986.

Cooney, James T., Regina Almeida, and Patricia Charles. "An Inter-American Community Education Network." COMMUNITY EDUCATION JOURNAL, X, No. 3, April 1983.

Goldsmith, Joanne. "The Challenge." Community Education Journal, IX, No. 4, July 1982.

Heynemann, Stephen. "Textbooks Do Make a Difference." WORLD BANK NEWS, II, No. 21, 1983.

Hueg, William F. Jr. and Craig A. Gannon, Eds. "Seminar Report: "Transforming Knowledge into Food in a Worldwide Context." University of Minnesota: Miller Publishing Company, 1978.

Malassis, Louis. THE RURAL WORLD. London and Paris: The UNESCO Press, 1976.

Meaders, O. Donald, et al. AGRICULTURAL EDUCATION IN AFRICA: A SYSTEM VIEW. Michigan State University, Agricultural and Extension Education and Institute of International Agriculture, September 1988.

OUR WORLD. Washington, D.C.: National Geographic Society, 1990.

UNESCO. STATISTICAL YEARBOOK, 1987. Paris, France: UNESCO, 1987.

United Nations Education, Scientific, and Cultural Organization. AGRICULTURAL EDUCATION IN ASIA: A REGIONAL SURVEY. Paris, France: Imprimerie Firmin-Didot, 1971.

Warnat, Winifred I. "World Class Work Force." VOCATIONAL EDUCATION JOURNAL, Vol. 66, No. 5, May 1991.

Welton, Richard et al. AGRICULTURAL EDUCATION IN PARAGUAY. NACTA JOURNAL, Vol. XXXI, No. 1, March 1987.

CHAPTER 14

"Agency Emphasizes Women in Development." USAID HIGHLIGHTS, Vol. 7, No. 3. Washington, D.C.: U.S. Agency for International Development, 1990.

Castillo, Gelia T. "The Farmer Revisited" PROCEEDINGS OF THE WORLD FOOD CONFERENCE. Ames, Iowa: Iowa State University Press, 1977.

"Cerescope" (Editorial). CERES: FAO REVIEW ON AGRICULTURE AND DEVELOPMENT, 14, No. 4. Rome, Italy: FAO, 1981.

Ebeling, Walter. THE FRUITED PLAIN. Berkeley, California: University of California Press, 1979.

Swanson, Burton E. AGRICULTURAL EXTENSION: A REFERENCE MANUAL, Second Edition. Rome, Italy: FAO, 1984.

CHAPTER 15

Agricultural Research Policy Advisory Committee. "Research to Meet U.S. and World Food Needs." Kansas City, Missouri: July 1975.
Boroson, J.E. and Warren Eberstadt. "The International Food Policy Research Institute." RF ILLUSTRATED, 4, No. 3. Washington, D.C.: Rockefeller Foundation, 1979.
Brochure. Consultative Group on International Agricultural Research, 1991.
Cummings, Ralph Jr. "Food Crops in Low Income Countries: The State of Present and Expected Agricultural Research and Technology." ROCKEFELLER FOUNDATION WORKING PAPERS. New York: 1980.
Hadwinger, Don F. "The Agricultural Research Establishment." Informal Consultation Report. Ames, Iowa: Iowa State University, 1980.
THE WORLD BANK ANNUAL REPORT, 1982. Washington, D.C.: The International Bank for Reconstruction and Development, 1982.
Wortman, Sterling and Ralph W. Cummings. TO FEED THIS WORLD. Baltimore: Johns Hopkins University Press, 1978.

CHAPTER 16

"Agricultural Cooperatives in Korea." Korean National Agricultural Cooperative Federation, 1991.
Bunch, Ronald. TWO EARS OF CORN: A GUIDE TO PEOPLE CENTERED AGRICULTURAL IMPROVEMENT. Oklahoma City, Oklahoma: World Neighbors, 1982.
Price, Robert R. "Ducks, Silkworms and Babies—Report on Consultative Tours to Bangladesh, Papua New Guinea, and Haiti." United Methodist Committee on Relief, Board of Global Ministries, United Methodist Church, September 1979.
Price, Robert R. "Recommendations: Relief and Agricultural Development at Lagonabo, Haiti", Board of Global Ministries, United Methodist Church, October 1977.
Reich, Otto J. "Latin America: The Economic and Political Challenge." AGENDA, 5, No. 4. Washington, D.C.: Agency for International Development, 1982.
Wennergren, E. Boyd and William Furlong. THE UNITED STATES AND WORLD AGRICULTURAL DEVELOPMENT. National Association of State Colleges and Land Grant Universities. Washington, D.C.: 1985.
Wortman, Sterling and Ralph W. Cummings. TO FEED THIS WORLD. Baltimore: Johns Hopkins University Press, 1978.

CHAPTER 17

Atala, T.K. "The Relationship of Socio-Economic Factors in Agricultural Innovation and Utilization of Information in Two Nigerian Villages." NIGERIAN JOURNAL OF AGRICULTURAL EXTENSION, 2(2), 1984.
Bunch, Roland. "The Alchemy of Success." NEW INTERNATIONALIST, No. 106. Leicester, U.K.: Blackfriars Press, December 1981.
Cairncross, J. POPULATION AND AGRICULTURE IN THE DEVELOPING COUNTRIES. FAO Development Paper. Rome, Italy: 1980.

Diamond, James E. "Philosophy of International Agricultural Development." NACTA JOURNAL, Vol. XXVIII, No. 4, December 1984

MacKinnon, A.R. "Education in Development." DEVELOPMENT. Canadian International Development Agency, Winter 1985.

Minear, Larry. NEW HOPE FOR THE HUNGRY. New York: Friendship Press, 1975.

Mushakoji, Kinhide. "Teaching for the Times." DEVELOPMENT FORUM, IX, No. 4. Geneva, Switzerland: United Nations University, May 1981.

Okunye, P.A. "Farmer Contact and the Demonstration of Important Farming Practices in Nigeria." AGRICULTURAL SYSTEMS, 18(4), 1985.

Onazi, O.C. "The Role of the Decision of Agricultural and Livestock Training Services Training in Training Intermediate-Level Manpower for Livestock Development in the Northern States of Nigeria". SAMARU AGRICULTURAL NEWSLETTER, 18 (1), 1976.

PROJECT TO INFUSE INTERNATIONAL AGRICULTURE INTO THE AGRICULTURAL EDUCATION CURRICULUM. Memorandum. The Council. Alexandria, Virginia: September 1989.

Sewell, John W. "Which Costs More: Aid or No Aid?" AGENDA, 4, No. 2. Washington, D.C.: Agency for International Development, 1981.

Stoskopf, L.P. et al. "Agricultural Education Developments in Northeast China." NACTA JOURNAL. Vol. XXX, No. 3, September 1986.

Toton, Suzanne C. WORLD HUNGER, THE RESPONSIBILITY OF CHRISTIAN EDUCATION. Maryknoll, New York: Orbis Books, 1982.

Tweeten, Luther. "The Hard (And Sometimes Hopeful) Facts about This Hungry World." WORLD VIEW, 21, No. 12, December 1978.

UNDERSTANDING AGRICULTURE—NEW DIRECTIONS FOR EDUCATION. Committee on Agricultural Education in Secondary Schools, Board of Agriculture, National Research Council. Washington, D.C.: National Academy Press, 1988.

UNITED STATES AGENCY FOR INTERNATIONAL DEVELOPMENT. African Bureau Agricultural Education Strategy Paper. Washington, D.C.: USAID, 1984.

World Bank Research Program. ABSTRACTS OF CURRENT STUDIES 1982. Washington, D.C.: World Bank, 1983.

Wortman, Sterling and Ralph W. Cummings. TO FEED THIS WORLD. Baltimore: Johns Hopkins University Press, 1978.

Zhang, Chi and Barbara Holt. "Agricultural Teacher Education Programs in China." NACTA JOURNAL. Vol. XXXIII, No. 2, June 1989.

Index

Adoption process, 200
  stage 1: awareness, 201–202
  stage 2: interest, 202–203
  stage 3: evaluation, 203
  stage 4: trial, 203
  stage 5: adoption, 203
Africa, 18
  food policy in, 154–56
Agency for International Development.
  See AID
Agrarian societies, 25
Agribusiness, world trade and, 143–45
Agricultural and food policy, 8, 153
Agricultural commodities, 129–49
Agricultural development, 13–14
  definition of, 258–59
  features of successful programs, 260–62
  importance of, 226–27
  local cooperatives as function of, 262–63
  program planning for, 259–60
Agricultural education
  definitions of, 277–78
  international, definition of, 11
  in other countries, 212–13, 217–20
  in Paraguay, 216–17
  role in agricultural development, 278–80
  in United States, 211–12
Agricultural extension
  elements conducive to effectiveness of, 238–40
  international, definition of, 12
  legislation needed for, 240–41
  need for rural youth in, 238
  resources needed for, 241
  teaching methods for, 241–45

Agriculturalists, 14–16
Agricultural research
  classification of, 254
  nature and scope of in developing countries, 254
  in other countries, 250–54
  in United States, 248–50
AID, 166, 173–74, 190–91
Aid programs, 166
  agencies and institutions providing, 167–72
  concepts of, 166
  government assistance programs, 172
  other organizations for, 178–79
  types of, 166
  United States programs, 173–76
  volunteer agencies for, 177–78
  See also Financial assistance
Albania, 145
Ancillary services, 239
Angola, 156
Animal products, 133
Applied technology, 111
Appropriate technology, 18, 109–23
Aquaculture, 73
Asian Development Bank, 188
Australia, farms in, 57

Bananas, 132
Bangladesh, 17–18
  farms in, 60
Basic research, 254
Beef, 72, 133
Behavioro-educationist, 267
BIFAD, 175–76
Bilateral agreement, 136
Biological research, 7
Biomass, 96

Board for International Food and Agricultural Development. *See* BIFAD
Brazil, farms in, 59
Bulgaria, change in agricultural policy in, 154

Canada
  exports of, 131
  farms in, 56–57
Canadian Wheat Board. *See* CWB
CAP, 137
CARE, 179
Cereal grains, 67, 69–70
CGIAR, 250–54
Change-agent, 199–200
Church assistance programs, 178–79
CIMMYT, 251
Citrus juices, 132–33
CMEA, 138
Collective farming systems, 66
Commercial banks (U.S.), 192
Commercial farming systems, 66
Commodities, 129–49
Common Agricultural Policy. *See* CAP
Commonwealth of Independent States, farms in, 57–58. *See also* Soviet Union
Communication media, 242
Community resource development, 259
Comparative advantage, 125–26
Competition, international, 145–47
Concessional, 136
Conferences, 244–45
Consultative Group on International Agricultural Research. *See* CGIAR
Consulting, 302–306
Consumer-ready products, 145
Controlling, 148
Cooperative Extension Service, 228–29
Cooperatives, 262–63
Cooperators, 229
Corporate farms, 158

Cotton, 133
Council of Mutual Economic Assistance. *See* CMEA
Crops
  cereal grains, 67, 69–70
  distribution and extent of, 66–67
  non-cereal, 70
Cultural attitudes, 26
Culture, food in, 44–45
CWB, 137

Dairy products, 133–34
Deforestation, 82
Demographic transition, 26–27
Denmark, farms in, 53–54
Desertification, 84–85
Developed countries, development education for, 220–21
Developing countries, 7
  agricultural research in, 250–54
  assistance programs for, 166–69
  disadvantages of, in trade and development, 269–70
  education in, 214–20
  energy consumption by food systems in, 97
  lack of commitment to agriculture in, 271–75
  low energy efficiency and output in, 98–99
  power sources in, 97–98
  trade disadvantages in, 138–39
  United States policies of assistance for, 160–61, 173–76
  women in, 231–37
Development aid, 166
Developmental assistance, 16–19
Development banks (U.S.), 191–92
Directing, 148
Dual economies, 221
Dutch Polders, 55–56

Ecodevelopment, 109

Ecological refugees, 81
Economic aid, 166
Economic development, 16
Economo-realist, 267
Education
  definition of, 196
  in developing nations, 214–16
  general vs. vocational, 210–11
  societal influences on, 221–23
  value of, for work, 211–12
  See also Agricultural education
EEC, 137
Embargoes, 135
Emergency aid, 166
Emerging countries, 7. *See also* Developing countries
Energy production, 88
Energy sources
  alternative, 96–97, 101
  fossil fuels, 95–96
  organic and inorganic, 95
Energy subsidy, 97
Enrollment ratio, 215
Esteem needs, 199
European Development Fund, 189
European Economic Community. *See* EEC
Experience, international. *See* International experience
Export-Import Bank, 190
Exports, 129–35
Extension field workers, 231

Family farms, 158
Family planning, 27
FAO, 168
Farming systems
  classification of, 63
  collective, 66
  commercial, 66
  corporate, 158
  definition of, 62
  family, 158
  group, 158
  influences on, 64–65
  state, 158
  subsistence, 65
Farm-level operational research, 254
Farm types, 52–60
  in Australia, 57
  in Bangladesh, 60
  in Brazil, 59
  in Canada, 56–57
  in Denmark, 53–54
  in Soviet Union, 57–58
  in France, 56
  in Haiti, 58–59
  in Kenya, 59–60
  in Netherlands, 55–56
  in Nigeria, 59
  in Norway, 54
  in South Korea, 57
  in Sweden, 55
Feed grains, 131
Financial assistance
  international programs for, 185–89
  lending institutions in United States for, 190–92
  for rural areas, 183–84
  United States foreign, 189
First World, 10
Fish, 73, 133
Food
  acquisition and distribution of, 156
  in a culture, 44–45
  community supply of, 40
  insecurity, 268–69
  production limitations of, 75–77, 90
  quality of, 40
  quantity of, 40
  timing of, 40
  world problem of, 37
Food aid, 166
Food and Agriculture Organization, 168
Food deficits, 25–26
Food for Peace Program, 18, 136, 174–75

Food trade, 126–28
Foreign assistance. *See* Aid
Foreign Assistance Act. *See* Title XII
Forests, 82–84
Fossil fuels, 95–96
Fragile environments, 81–82
France, farms in, 56
Functional education, 211
Future-oriented societies, 210

GATT, 137
General Agreement of Trade and Tariffs. *See* GATT
General education, 210
Global agriculture, 8, 10
  environmental/social influences on, 80–90
  population/consumption explosion and, 23
Global responsibility, 36
Gross national product, 107
Group farms, 158

Haiti, farms in, 58–59
HEART program, 111–15
Hectare, 61
Heifer Project International, 115–17
Horticultural products, 132–33
Humanistic-Democratic-Participative systems, 205
Human resource development
  adoption process, 200–204
  change-agent as teacher, 199–200
  constructive participation, maintenance of, 204–205
  education for, 196–97
  incentives, maintenance of, 204
  learning and motivation for, 195–96
  Maslow's theory of hierarchical needs and, 197–99
  outstanding trainers, 205–206
Human resources, 65
Hunger, world, 37–38

Hunger cycle, 41–42
Hydropower, 98

IBRD, 187
IDA, 187
IDB, 188
Idle land, 45
IESC, 177
IFPRI, 252, 254
Illiteracy, 42, 44, 270–71
IMF, 188
Indonesia, 145
Industrial crops, 59
Industrial roundwood, 83
Infant mortality, 38
Infrastructure, 64
Inorganic energy sources, 95
Intellectual investment, 211
Interagency Staff Committee. *See* ISC
Inter-American Development Bank. *See* IDB
International agricultural extension, 11–13
  organization of, 229–30
  problems of, 230–31
  programs for United States citizens, 311–15
International Agricultural Research Institutes, 251–54
International Bank for Reconstruction and Development. *See* IBRD
International Centre for the Improvement of Maize and Wheat. *See* CIMMYT
International Development Association. *See* IDA
International Executive Service Corps. *See* IESC
International experience, gaining, 282–83
  considerations for, 283
  foundations/private organizations for, 291–94

international associations for, 296–97
international directories for, 297–300
international magazines for, 300–301
international organizations for, 295–96
programs for, 284–88
publications for, 288
suggestions for, 284
United States government organizations for, 289–90
International Food Policy Research Institute. *See* IFPRI
International Learning and Livestock Center, 115–17
International Monetary Fund. *See* IMF
International Rice Research Institute. *See* IRRI
IRRI, 251
ISC, 136

Japan, 33, 144

Kenya, farms in, 59–60
Korea, cooperatives in, 263

Labor-intensive economies, 99
Lamb, 133
Land Grant University, 177–78, 219–20, 248, 316–18
international role of, 316–21
resident instruction programs, 318–19
Land tenure, 158
Learning
definition of, 196
motivation and, 195–96
stages of, 200–203
Twelve Principles of, 197
Lectures, 244
Levies, 137
Liberia, 156
Literacy levels, 215–16

Livestock, 70–73
Local cooperatives, 262–63

Machino-technologist, 267
Malnutrition
definition of, 38–39
factors affecting, 40, 42–45
maternal depletion syndrome and, 40–41
steps in overcoming, 45–50
See also Nutritional problems; World hunger
Management functions, 147–49
Marketing, 8
Maslow's Theory, 197–99
Mass media, 245
Maternal depletion syndrome, 40–41
Meat animals, 71–72
Meetings, 244–45
Method demonstration, 243–44
Metropolitan areas, 32–33
Mexico, 33
Microentrepreneurs, 233
Migration, to metropolitan areas, 32–33
Military aid, 166
Milk, 73–74
Ministry of Agriculture, 229–30
Mozambique, 156
Multilateral agreement, 136
Mutton, 133

National Association of State Universities and Land Grant Colleges basic principles for international involvement, 322–25
Natural resources, 64
Netherlands, farms in, 55–56
Nigeria, farms in, 59
Non-cereal crops, 70
Nontariff restrictions, 136
Norway, farms in, 54
Nutritional problems
causes of, 268–75

Nutritional problems (*Cont.*)
persistence of, 266–67
See also Malnutrition; World hunger

Oilseeds, 131
Optimal environmental conditions, 81
task of maintaining, 89
Organic energy sources, 95
Organizing, 147
Overpopulation, definition of, 24

Paraguay, agricultural education in, 216–17
Participative program planning, 259–60
Peace Corps, 177
People's Republic of China, 131
Planning, 147
Poland, 145
change in agricultural policy in, 153–54
Political, definition of, 152
Political instability, 269
Politico-parochialist, 267
Politics, definition of, 152
Pollution, 88–89
Population growth, 23–24
factors and policies in limiting, 26–27
migration to metropolitan areas and, 32–33
nature of, in agrarian societies, 25
projection figures by nation (year 2100), 27–32
rates of, and food deficits, 25–26, 269
Pork, 133
Poultry, 133
Poverty, 42, 90
absolute, 268
relative, 268
Power sources, in developing nations, 97–98

Private Voluntary Organizations. *See* PVOs
Production/consumption comparisons, 49
Production incentives, 64–65
Productivity, range in, 60–62
Protectionism, 139
Public Law 480, 136, 174–75
Public policy, 8, 65
Pupil/teacher ratio, 215
PVOs, 179

Religious organizations, assistance programs of, 178–79
Research, agricultural. *See* Agricultural research
Resources, 64
use and allocation of, 159
Result demonstration, 243
Rice, 131
Ruminant animals, 72
Rural development, definition of, 259
Rural financial assistance, 183–84
Rural village centers, 160
Rural young people, 238

Safety needs, 197
Salinization, 85, 87
Second World, 10
Sedimentation, 85, 87
Self-actualization, 199
Slash-and-burn agriculture, 84
Small-scale experimentation, 260–61
Smith-Lever Act, 229
Social needs, 199
Socio/religio-humanist, 267
South Korea, farms in, 57
Soviet Union, food crisis in, 307–10
See also Commonwealth of Independent States
Soybeans, 131
Staffing, 148
State farms, 158

State Universities, 177–78, 219–20
Strategic research, 254
Subsidies, 137
Subsistence farming systems, 65
Supporting research, 254
Support systems, 159
Survival needs, 197
Sustainable agriculture, 90
Sweden, farms in, 55
Swine, 72–73

Tactical research, 254
Tariffs, 135
Teaching
  methods for, 241–45
  principles of, 199–200
Technology, 64
  applied, 111
  appropriate, 8, 109–23
  definition of, 104
  economic growth and stability and, 107–108
  food production and, 104–107
  organization for rural development and, 108–109
Technology transfer, 5, 107, 195
  criteria for, 109–11
  slow pace of, 45
Thailand, 131
Third World country, 10. *See also* Developing countries
Title XII (Foreign Assistance Act)
  BIFAD and, 175–76
  programs in developing countries, 176
  programs in United States, 176
Tours, 245
Trade
  agreements between nations, 136–38
  benefits of, 138
  comparative advantage theory and, 125–26
  competition in world market, 145–47
  disadvantages of, 138–39
  in excess foodstuffs, 128–29
  food trade, importance of, 126–28
  management functions basic to, 147–49
  restriction policies in, 135–36
  types of commodities traded, 129–35
  United States and, 140–43
Trade concessions, 135–36
Traditional societies, 210
Trainers, outstanding, 205–206
Transfer pricing, 136
Transnational parent corporation, 136
Transportation, 88

Underdeveloped countries, 7. *See also* Developing countries
UNDP, 168–72, 185–87
Unilateral agreement, 136
United Nations Development Program. *See* UNDP
United States
  agricultural research in, 248–50
  agricultural status in, 4–8
  agricultural world trade and, 140–43
  assistance programs of, 173–76
  commodities traded by, 129
  foreign assistance, 189
  lending institutions in for foreign assistance, 190–92
  policies for developing nations, 160–61
  Title XII programs in, 176
  vocational education in, 211–12
University assistance programs, 177–78, 219–20
Urbanization, 88
USAID, 174
USDA/state research program, 249

Value-added, 145

Vocational education, 210
Volunteer agencies, 177–78

Waterlogging, 85, 87
Weather, 64
Wheat, 130–31
Winrock International, 117–23
Women
　in developing world, 237
　as farmers and laborers, 231–37
　in United States who farm, 237
World agricultural demand, 74, 77f
World Bank, 187–88
World food problem, definition of, 37
World hunger
　causes of, 42–45
　definition of, 38
　See also Malnutrition; Nutritional problems